The Very Basics of Tensors

The Very Basics of Tensors

Nils K. Oeijord

iUniverse, Inc.
New York Lincoln Shanghai

The Very Basics of Tensors

iUniverse books may be ordered through booksellers or by contacting:

iUniverse
2021 Pine Lake Road, Suite 100
Lincoln, NE 68512
www.iuniverse.com
1-800-Authors (1-800-288-4677)

ISBN-13: 978-0-595-35694-2 (pbk)
ISBN-13: 978-0-595-80172-5 (ebk)
ISBN-10: 0-595-35694-X (pbk)
ISBN-10: 0-595-80172-2 (ebk)

Printed in the United States of America

Contents

Preface

Tensor calculus is one of the main instruments of modern mathematics. The word "tensor" stems from the Latin word "tensus" meaning stretched. Actually, tensors were first utilized to describe the elastic deformation of solids. Tensor calculus is a generalization of vector calculus. It was developed by the Italian mathematicians Ricci-Curbastro and Levi Civita before 1900, but the basic ideas of tensor calculus originated with Riemann. Tensor calculus came into prominence in 1916 with the publication of the general theory of relativity. Tensor calculus comes near to being a universal language in physics. Tensor calculus is important in differential geometry, Riemannian geometry, mechanics of continua, and the general theory of relativity. It provides the only suitable mathematical language for general discussions in the general theory of relativity. Tensor calculus can also be used in the special theory of relativity, Euclidean geometry, classical dynamics, hydrodynamics, electromagnetics, and numerous other fields of science and engineering. Physical laws must be independent of any particular coordinate system used in describing them. This requirement leads to tensor calculus. There are two distinct approaches to the teaching of tensors: the conventional approach based on indices and the index-free approach. The main advantage of the index-free approach is that it offers deeper geometrical insight. The main disadvantage of this approach is that when one wants to do a real calculation with tensors, then recourse has to be made to the conventional approach. We shall adopt the conventional approach. The only prerequisites for reading this book are a familiarity with calculus (including vector calculus) and linear algebra, and some knowledge of differential equations.

1. Spaces, points, and coordinates

It is well known that in a 2-space (two dimensional space) and in a 3-space (three dimensional space) a *point* is a set of two and three numbers, respectively, for example (x, y) and (x, y, z), called *coordinates.*

By analogy, a set of N numbers, for example $\left(x^1, x^2, \ldots, x^N\right)$, where 1, 2,…, N are superscripts, not exponents, is a point in a kind of N-space with a kind of coordinate system.

Of course, the fact that we cannot visualize points in spaces of dimension higher than three, has nothing to do with their mathematical, and even physical, existence.

2. Distance between adjacent points in a Euclidean 2-space (for example an ordinary plane), a Euclidean 3-space (for example ordinary space), and a Euclidean *N*-space

Let (x, y) and $(x + dx, y + dy)$ be the rectangular Cartesian coordinates of two adjacent points in ordinary Euclidean 2-space, and let (x, y, z) and $(x + dx, y + dy, z + dz)$ be the rectangular Cartesian coordinates of two adjacent points in ordinary Euclidean 3-space. Then it is well known that the square of the distance between two adjacent points is

$$ds^2 = dx^2 + dy^2$$

$$ds^2 = dx^2 + dy^2 + dz^2$$

respectively. By analogy, in a particular *N*-space: if it is possible to find a particular coordinate system $(x^1, x^2, \ldots, x^N)$ such that $ds^2 = (dx^1)^2 + (dx^2)^2 + \ldots + (dx^N)^2$, then this particular *N*-space is, by definition, an Euclidean *N*-space.

We will learn that, the value of ds^2 is independent of the coordinate system used. Such a quantity is called an *invariant* or a *tensor of rank zero*. Tensor calculus works in a general *differentiable* space of N dimensions, where $N \geq 2$. Note that tensor calculus can work in a *non-metrical space*, in which the concept of ds^2 never enters.

3. Flat and curved spaces

It has been proved that the angles of an "ordinary" triangle in a Euclidean N-space sum to 180 degrees. Therefore Euclidean spaces are said to be *flat.* Examples: The surfaces of an ordinary cylinder and an ordinary cone are Euclidean 2-spaces, and are flat in the sense of our "definition". But the angles of a triangle on the surface of a sphere do not sum to 180 degrees, and therefore the surface of the sphere is not flat. A space which is not flat is called *curved.* See more on flat and curved spaces later in this book.

4. Transformation of coordinates in Euclidean 3-space. Curvilinear coordinates

Let the space be Euclidean 3-space and let the rectangular Cartesian coordinates (*x, y, z*) of any point in this space be expressed as functions of (u^1, u^2, u^3) so that

$$x = x(u^1, u^2, u^3)$$
$$y = y(u^1, u^2, u^3)$$
$$z = z(u^1, u^2, u^3)$$

Note that superscripts are not exponents in tensor notation. For reasons which will appear later, it is best to write the numerical labels as superscripts rather than subscripts. Suppose that the equations above can be solved for u^1, u^2, u^3 in terms of *x, y, z*. Then

$$u^1 = u^1(x, y, z)$$
$$u^2 = u^2(x, y, z)$$
$$u^3 = u^3(x, y, z)$$

Assume that the correspondence between (*x, y, z*) and (u^1, u^2, u^3) is unique, and that the functions x, y, z, u^1, u^2, u^3 have continuous derivatives. If these assumptions do not apply at certain points, then special consideration is required. A point *P* with rectangular Cartesian coordinates (*x, y, z*) has *curvilinear coordinates* (u^1, u^2, u^3). The equations above define a *transformation of coordinates.* Of course, a rectangular Cartesian *coordinate system* is only a kind of *general curvilinear coordinate systems.*

5. Orthogonal curvilinear coordinates

Let us in Euclidean 3-space take two coordinate systems: A rectangular Cartesian coordinate system (x, y, z) and a more general curvilinear coordinate system (u^1, u^2, u^3). The surfaces

$$u^1(x, y, z) = c_1$$
$$u^2(x, y, z) = c_2$$
$$u^3(x, y, z) = c_3$$

where c_1, c_2, c_3 are constants, are called *coordinate surfaces.* Each pair of the coordinate surfaces intersect in curves called *coordinate curves.* If the coordinate surfaces intersect at right angles the curvilinear coordinate system is called an *orthogonal curvilinear coordinate system.* The u^1, u^2, u^3 curves are, of course, analogous to the *x, y, z* axes of a rectangular Cartesian coordinate system.

6. Vectors in Euclidean 3-space with general curvilinear coordinates

The *position vector* $\vec{r}(u^1, u^2, u^3)$ of a point P (u^1, u^2, u^3) or P (x, y, z), in Euclidean 3-space with general curvilinear coordinates (u^1, u^2, u^3) and rectangular Cartesian coordinates (x, y, z) can be written

$$\vec{r}(u^1, u^2, u^3) = x(u^1, u^2, u^3)\vec{i} + y(u^1, u^2, u^3)\vec{j} + z(u^1, u^2, u^3)\vec{k}$$

where $\vec{i}, \vec{j}, \vec{k}$ are the well known $\vec{i}, \vec{j}, \vec{k}$ unit vectors in rectangular Cartesian coordinates in Euclidean 3-space. From vector calculus or vector analysis we know that the *tangent vectors* to the u^1, u^2, u^3 curves at P (u^1, u^2, u^3) in the directions of increasing u^1, u^2, u^3 are, respectively,

$$\frac{\partial \vec{r}(u^1, u^2, u^3)}{\partial u^1} = \frac{\partial x(u^1, u^2, u^3)}{\partial u^1}\vec{i} + \frac{\partial y(u^1, u^2, u^3)}{\partial u^1}\vec{j} + \frac{\partial z(u^1, u^2, u^3)}{\partial u^1}\vec{k}$$

$$\frac{\partial \vec{r}(u^1, u^2, u^3)}{\partial u^2} = \frac{\partial x(u^1, u^2, u^3)}{\partial u^2}\vec{i} + \frac{\partial y(u^1, u^2, u^3)}{\partial u^2}\vec{j} + \frac{\partial z(u^1, u^2, u^3)}{\partial u^2}\vec{k}$$

$$\frac{\partial \vec{r}(u^1, u^2, u^3)}{\partial u^3} = \frac{\partial x(u^1, u^2, u^3)}{\partial u^3}\vec{i} + \frac{\partial y(u^1, u^2, u^3)}{\partial u^3}\vec{j} + \frac{\partial z(u^1, u^2, u^3)}{\partial u^3}\vec{k}$$

Then the *unit tangent vectors* at P to the u^1, u^2, u^3 curves in the directions of increasing u^1, u^2, u^3 are, respectively,

$$\vec{e}_1 = \frac{\dfrac{\partial \vec{r}}{\partial u^1}}{\left|\dfrac{\partial \vec{r}}{\partial u^1}\right|}, \qquad \vec{e}_2 = \frac{\dfrac{\partial \vec{r}}{\partial u^2}}{\left|\dfrac{\partial \vec{r}}{\partial u^2}\right|}, \qquad \vec{e}_3 = \frac{\dfrac{\partial \vec{r}}{\partial u^3}}{\left|\dfrac{\partial \vec{r}}{\partial u^3}\right|}$$

If we put

$$h_1 = \left|\frac{\partial \vec{r}}{\partial u^1}\right|, \quad h_2 = \left|\frac{\partial \vec{r}}{\partial u^2}\right|, \quad h_3 = \left|\frac{\partial \vec{r}}{\partial u^3}\right|$$

we have

$$\frac{\partial \vec{r}}{\partial u^1} = h_1\, \vec{e}_1, \quad \frac{\partial \vec{r}}{\partial u^2} = h_2\, \vec{e}_2, \quad \frac{\partial \vec{r}}{\partial u^3} = h_3\, \vec{e}_3$$

The factors h_1, h_2, h_3 are called *scale factors*. The $\vec{e}_1, \vec{e}_2, \vec{e}_3$ vectors are, of course, analogous to the $\vec{i}, \vec{j}, \vec{k}$ vectors of a rectangular Cartesian coordinate system (in Euclidean 3-space). From calculus we have

$$h_1 = \left|\frac{\partial \vec{r}}{\partial u^1}\right| = \sqrt{\left(\frac{\partial x}{\partial u^1}\right)^2 + \left(\frac{\partial y}{\partial u^1}\right)^2 + \left(\frac{\partial z}{\partial u^1}\right)^2}$$

$$h_2 = \left|\frac{\partial \vec{r}}{\partial u^2}\right| = \sqrt{\left(\frac{\partial x}{\partial u^2}\right)^2 + \left(\frac{\partial y}{\partial u^2}\right)^2 + \left(\frac{\partial z}{\partial u^2}\right)^2}$$

$$h_3 = \left|\frac{\partial \vec{r}}{\partial u^3}\right| = \sqrt{\left(\frac{\partial x}{\partial u^3}\right)^2 + \left(\frac{\partial y}{\partial u^3}\right)^2 + \left(\frac{\partial z}{\partial u^3}\right)^2}$$

or

$$h_1^2 = \left(\frac{\partial x}{\partial u^1}\right)^2 + \left(\frac{\partial y}{\partial u^1}\right)^2 + \left(\frac{\partial z}{\partial u^1}\right)^2$$

$$h_2^2 = \left(\frac{\partial x}{\partial u^2}\right)^2 + \left(\frac{\partial y}{\partial u^2}\right)^2 + \left(\frac{\partial z}{\partial u^2}\right)^2$$

$$h_3^2 = \left(\frac{\partial x}{\partial u^3}\right)^2 + \left(\frac{\partial y}{\partial u^3}\right)^2 + \left(\frac{\partial z}{\partial u^3}\right)^2$$

Also, from calculus:

$$\nabla u^1(x,y,z) = \frac{\partial u^1}{\partial x}\vec{i} + \frac{\partial u^1}{\partial y}\vec{j} + \frac{\partial u^1}{\partial z}\vec{k}$$
$$\nabla u^2(x,y,z) = \frac{\partial u^2}{\partial x}\vec{i} + \frac{\partial u^2}{\partial y}\vec{j} + \frac{\partial u^2}{\partial z}\vec{k}$$
$$\nabla u^3(x,y,z) = \frac{\partial u^3}{\partial x}\vec{i} + \frac{\partial u^3}{\partial y}\vec{j} + \frac{\partial u^3}{\partial z}\vec{k}$$

are vectors at P normal to the coordinate surfaces

$$u^1(x,y,z) = c_1$$
$$u^2(x,y,z) = c_2$$
$$u^3(x,y,z) = c_3$$

respectively, where c_1, c_2, c_3 are constants and the coordinates of P is (x, y, z) in rectangular Cartesian coordinates. The unit vectors at P in the directions of

$$\nabla u^1(x,y,z)$$
$$\nabla u^2(x,y,z)$$
$$\nabla u^3(x,y,z)$$

are, respectively

$$\vec{E}_1 = \frac{\nabla u^1}{|\nabla u^1|}$$
$$\vec{E}_2 = \frac{\nabla u^2}{|\nabla u^2|}$$
$$\vec{E}_3 = \frac{\nabla u^3}{|\nabla u^3|}$$

$\vec{E}_1, \vec{E}_2, \vec{E}_3$, are unit vectors at P normal to the coordinate surfaces

$$u^1(x, y, z) = c_1$$
$$u^2(x, y, z) = c_2$$
$$u^3(x, y, z) = c_3$$

respectively. Thus at each point P of a curvilinear coordinate system in Euclidean 3-space there exist two sets of unit vectors, $\vec{e}_1, \vec{e}_2, \vec{e}_3$, and $\vec{E}_1, \vec{E}_2, \vec{E}_3$ Both sets are analogous to the $\vec{i}, \vec{j}, \vec{k}$ vectors in rectangular Cartesian coordinates, but in general $\vec{e}_1, \vec{e}_2, \vec{e}_3$ and $\vec{E}_1, \vec{E}_2, \vec{E}_3$ change directions from point to point. The sets $\vec{e}_1, \vec{e}_2, \vec{e}_3$ and $\vec{E}_1, \vec{E}_2, \vec{E}_3$ become identical sets if and only if the curvilinear coordinate system is orthogonal.

7. Contravariant and covariant components of a vector in Euclidean 3-space

A vector $\vec{A}$ in Euclidean 3-space with a curvilinear coordinate system (u^1, u^2, u^3) can be represented in terms of the *unit base vectors* $\vec{e}_1, \vec{e}_2, \vec{e}_3$ or $\vec{E}_1, \vec{E}_2, \vec{E}_3$:

$$\vec{A} = A_1 \, \vec{e}_1 + A_2 \, \vec{e}_2 + A_3 \, \vec{e}_3$$

$$\vec{A} = a_1 \, \vec{E}_1 + a_2 \, \vec{E}_2 + a_3 \, \vec{E}_3$$

A_1, A_2, A_3 and a_1, a_2, a_3 are called the *components* of $\vec{A}$ in each system, respectively. We can, of course, use

$$\frac{\partial \vec{r}}{\partial u^1}, \frac{\partial \vec{r}}{\partial u^2}, \frac{\partial \vec{r}}{\partial u^3} \quad \text{or} \quad \nabla u^1, \ \nabla u^2, \ \nabla u^3$$

as base vectors (called *unitary base vectors*):

$$\vec{A} = C^1 \frac{\partial \vec{r}}{\partial u^1} + C^2 \frac{\partial \vec{r}}{\partial u^2} + C^3 \frac{\partial \vec{r}}{\partial u^3}$$

$$\vec{A} = C_1 \, \nabla u^1 + C_2 \, \nabla u^2 + C_3 \, \nabla u^3$$

The unitary base vectors *are not* unit vectors in general. C^1, C^2, C^3 are called the *contravariant components* of $\vec{A}$, and C_1, C_2, C_3 are called the *covariant components* of $\vec{A}$. For contravariant components we use superscripts instead of subscripts, and for covariant components we use subscripts, a policy which will prove useful.

8. The relation between the contravariant components of a vector in two general curvilinear coordinate systems in Euclidean 3-space

Let $\vec{A}$ be an arbitrary vector in Euclidean 3-space with a rectangular Cartesian coordinate system $(x,\ y,\ z)$ and two general curvilinear coordinate systems (u^1, u^2, u^3) and $(\bar{u}^1, \bar{u}^2, \bar{u}^3)$. If the transformation equations

$$x = x_1\,(u^1, u^2, u^3)$$

$$y = y_1(u^1, u^2, u^3)$$

$$z = z_1(u^1, u^2, u^3)$$

$$x = x_2(\bar{u}^1, \bar{u}^2, \bar{u}^3)$$

$$y = y_2(\bar{u}^1, \bar{u}^2, \bar{u}^3)$$

$$z = z_2(\bar{u}^1, \bar{u}^2, \bar{u}^3)$$

exist, then the transformation equations

$$u^1 = u^1(\bar{u}^1, \bar{u}^2, \bar{u}^3)$$

$$u^2 = u^2(\bar{u}^1, \bar{u}^2, \bar{u}^3)$$

$$u^3 = u^3(\bar{u}^1, \bar{u}^2, \bar{u}^3)$$

$$\bar{u}^1 = \bar{u}^1(u^1, u^2, u^3)$$

$$\bar{u}^2 = \bar{u}^2(u^1, u^2, u^3)$$

$$\bar{u}^3 = \bar{u}^3(u^1, u^2, u^3)$$

exist. Assume that the functions above have continuous derivatives. Then it can be proved (we shall do it in section 10) that

$$C^1 = \bar{C}^1 \frac{\partial u^1}{\partial \bar{u}^1} + \bar{C}^2 \frac{\partial u^1}{\partial \bar{u}^2} + \bar{C}^3 \frac{\partial u^1}{\partial \bar{u}^3}$$

$$C^2 = \bar{C}^1 \frac{\partial u^2}{\partial \bar{u}^1} + \bar{C}^2 \frac{\partial u^2}{\partial \bar{u}^2} + \bar{C}^3 \frac{\partial u^2}{\partial \bar{u}^3}$$

$$C^3 = \bar{C}^1 \frac{\partial u^3}{\partial \bar{u}^1} + \bar{C}^2 \frac{\partial u^3}{\partial \bar{u}^2} + \bar{C}^3 \frac{\partial u^3}{\partial \bar{u}^3}$$

or, in shorter notation,

$$C^p = \sum_{q=1}^{3} \bar{C}^q \frac{\partial u^p}{\partial \bar{u}^q} \qquad p = 1, 2, 3$$

An even shorter notation is simply to write

$$C^p = \bar{C}^q \frac{\partial u^p}{\partial \bar{u}^q}$$

or

$$C^p = \frac{\partial u^p}{\partial \bar{u}^q} \bar{C}^q$$

where we adopt the *summation convention* that whenever an index (superscript or subscript) is repeated in a given term we are to sum over that index (called a dummy index) from 1 to N, (N is the number of dimensions of the space), and the *range convention* that an index occurring only once in a given term (called a *free index*) can stand for any of the numbers 1, 2, …, N, where N is the number of dimensions of the space.

Similarly, by interchanging the coordinates, we can show that

$$\bar{C}^p = \frac{\partial \bar{u}^p}{\partial u^q} C^q$$

9. The relation between the covariant components of a vector in two general curvilinear coordinate systems in Euclidean 3-space

Let $\vec{A}$ be an arbitrary vector in Euclidean 3-space with a rectangular Cartesian coordinate system (x, y, z) and two general curvilinear coordinate systems (u^1, u^2, u^3) and $(\bar{u}^1, \bar{u}^2, \bar{u}^3)$. If the transformation equations

$$x = x_1\,(u^1, u^2, u^3)$$

$$y = y_1(u^1, u^2, u^3)$$

$$z = z_1(u^1, u^2, u^3)$$

$$x = x_2(\bar{u}^1, \bar{u}^2, \bar{u}^3)$$

$$y = y_2(\bar{u}^1, \bar{u}^2, \bar{u}^3)$$

$$z = z_2(\bar{u}^1, \bar{u}^2, \bar{u}^3)$$

exist, then the transformation equations

$$u^1 = u^1(\bar{u}^1, \bar{u}^2, \bar{u}^3)$$

$$u^2 = u^2(\bar{u}^1, \bar{u}^2, \bar{u}^3)$$

$$u^3 = u^3(\bar{u}^1, \bar{u}^2, \bar{u}^3)$$

$$\bar{u}^1 = \bar{u}^1(u^1, u^2, u^3)$$

$$\bar{u}^2 = \bar{u}^2(u^1, u^2, u^3)$$

$$\bar{u}^3 = \bar{u}^3(u^1, u^2, u^3)$$

exist. Assume that the functions above have continuous derivatives. Then it can be proved (we shall do it in section 11) that

$$C_1 = \bar{C}_1 \frac{\partial \bar{u}^1}{\partial u^1} + \bar{C}_2 \frac{\partial \bar{u}^2}{\partial u^1} + \bar{C}_3 \frac{\partial \bar{u}^3}{\partial u^1}$$

$$C_2 = \bar{C}_1 \frac{\partial \bar{u}^1}{\partial u^2} + \bar{C}_2 \frac{\partial \bar{u}^2}{\partial u^2} + \bar{C}_3 \frac{\partial \bar{u}^3}{\partial u^2}$$

$$C_3 = \bar{C}_1 \frac{\partial \bar{u}^1}{\partial u^3} + \bar{C}_2 \frac{\partial \bar{u}^2}{\partial u^3} + \bar{C}_3 \frac{\partial \bar{u}^3}{\partial u^3}$$

or, in shorter notation,

$$C_p = \sum_{q=1}^{3} \bar{C}_q \frac{\partial \bar{u}^q}{\partial u^p} \qquad p = 1, 2, 3$$

or (using the summation and range conventions)

$$C_p = \bar{C}_q \frac{\partial \bar{u}^q}{\partial u^p}$$

or

$$C_p = \frac{\partial \bar{u}^q}{\partial u^p} \bar{C}_q$$

Similarly, by interchanging the coordinates, we can show that

$$\bar{C}_p = \frac{\partial u^q}{\partial \bar{u}^p} C_q$$

10. Proof of the result in section 8. Generalization to *N*-space

The *position vector* $\vec{r}(u^1, u^2, u^3)$ of a point P (u^1, u^2, u^3) or P (x, y, z), in Euclidean 3-space with general curvilinear coordinates (u^1, u^2, u^3) and rectangular Cartesian coordinates (x, y, z) can be written

$$\vec{r}(u^1, u^2, u^3) = x(u^1, u^2, u^3)\vec{i} + y(u^1, u^2, u^3)\vec{j} + z(u^1, u^2, u^3)\vec{k}$$

Similarly, by interchanging the curvilinear coordinates, we see that

$$\vec{r}(\bar{u}^1, \bar{u}^2, \bar{u}^3) = x(\bar{u}^1, \bar{u}^2, \bar{u}^3)\vec{i} + y(\bar{u}^1, \bar{u}^2, \bar{u}^3)\vec{j} + z(\bar{u}^1, \bar{u}^2, \bar{u}^3)\vec{k}$$

Then, from calculus, we have

$$d\vec{r} = \frac{\partial \vec{r}}{\partial u^1} du^1 + \frac{\partial \vec{r}}{\partial u^2} du^2 + \frac{\partial \vec{r}}{\partial u^3} du^3 = \vec{\alpha}_1\, du^1 + \vec{\alpha}_2\, du^2 + \vec{\alpha}_3\, du^3$$

and

$$d\vec{r} = \frac{\partial \vec{r}}{\partial \bar{u}^1} d\bar{u}^1 + \frac{\partial \vec{r}}{\partial \bar{u}^2} d\bar{u}^2 + \frac{\partial \vec{r}}{\partial \bar{u}^3} d\bar{u}^3 = \bar{\vec{\alpha}}_1\, d\bar{u}^1 + \bar{\vec{\alpha}}_2\, d\bar{u}^2 + \bar{\vec{\alpha}}_3\, d\bar{u}^3$$

Then

$$\vec{\alpha}_1\, du^1 + \vec{\alpha}_2\, du^2 + \vec{\alpha}_3\, du^3 = \bar{\vec{\alpha}}_1\, d\bar{u}^1 + \bar{\vec{\alpha}}_2\, d\bar{u}^2 + \bar{\vec{\alpha}}_3\, d\bar{u}^3$$

We know from section 8 that

$$u^1 = u^1(\bar{u}^1, \bar{u}^2, \bar{u}^3)$$

$$u^2 = u^2(\bar{u}^1, \bar{u}^2, \bar{u}^3)$$

$$u^3 = u^3(\bar{u}^1, \bar{u}^2, \bar{u}^3)$$

Then, from calculus

$$
\begin{aligned}
du^1 &= \frac{\partial u^1}{\partial \bar{u}^1} d\bar{u}^1 + \frac{\partial u^1}{\partial \bar{u}^2} d\bar{u}^2 + \frac{\partial u^1}{\partial \bar{u}^3} d\bar{u}^3 \\
du^2 &= \frac{\partial u^2}{\partial \bar{u}^1} d\bar{u}^1 + \frac{\partial u^2}{\partial \bar{u}^2} d\bar{u}^2 + \frac{\partial u^2}{\partial \bar{u}^3} d\bar{u}^3 \\
du^3 &= \frac{\partial u^3}{\partial \bar{u}^1} d\bar{u}^1 + \frac{\partial u^3}{\partial \bar{u}^2} d\bar{u}^2 + \frac{\partial u^3}{\partial \bar{u}^3} d\bar{u}^3
\end{aligned}
$$

Substituting and equating coefficients, we find

$$
\begin{aligned}
\bar{\vec{\alpha}}_1 &= \vec{\alpha}_1 \frac{\partial u^1}{\partial u^1} + \vec{\alpha}_2 \frac{\partial u^2}{\partial u^1} + \vec{\alpha}_3 \frac{\partial u^3}{\partial u^1} \\
\bar{\vec{\alpha}}_2 &= \vec{\alpha}_1 \frac{\partial u^1}{\partial u^2} + \vec{\alpha}_2 \frac{\partial u^2}{\partial u^2} + \vec{\alpha}_3 \frac{\partial u^3}{\partial u^2} \\
\bar{\vec{\alpha}}_3 &= \vec{\alpha}_1 \frac{\partial u^1}{\partial u^3} + \vec{\alpha}_2 \frac{\partial u^2}{\partial u^3} + \vec{\alpha}_3 \frac{\partial u^3}{\partial u^3}
\end{aligned}
$$

Now a vector $\vec{A}$ can be expressed in the two curvilinear coordinate systems as

$$
\begin{aligned}
\vec{A} &= C^1 \vec{\alpha}_1 + C^2 \vec{\alpha}_2 + C^3 \vec{\alpha}_3 \\
\vec{A} &= \bar{C}^1 \bar{\vec{\alpha}}_1 + \bar{C}^2 \bar{\vec{\alpha}}_2 + \bar{C}^3 \bar{\vec{\alpha}}_3
\end{aligned}
$$

Equating equations, substituting, and equating coefficients, we find

$$
C^1 = \bar{C}^1 \frac{\partial u^1}{\partial \bar{u}^1} + \bar{C}^2 \frac{\partial u^1}{\partial \bar{u}^2} + \bar{C}^3 \frac{\partial u^1}{\partial \bar{u}^3}
$$

$$
C^2 = \bar{C}^1 \frac{\partial u^2}{\partial \bar{u}^1} + \bar{C}^2 \frac{\partial u^2}{\partial \bar{u}^2} + \bar{C}^3 \frac{\partial u^2}{\partial \bar{u}^3}
$$

$$
C^3 = \bar{C}^1 \frac{\partial u^3}{\partial \bar{u}^1} + \bar{C}^2 \frac{\partial u^3}{\partial \bar{u}^2} + \bar{C}^3 \frac{\partial u^3}{\partial \bar{u}^3}
$$

or

$$C^p = \overline{C}^q \frac{\partial u^p}{\partial \overline{u}^q}$$

or

$$C^p = \frac{\partial u^p}{\partial \overline{u}^q} \overline{C}^q$$

or, by interchanging the coordinates, we can show that

$$\overline{C}^p = \frac{\partial \overline{u}^p}{\partial u^q} C^q$$

The above results lead us to adopt the following definition: If N quantities $C^1, C^2, \ldots, C^N$ of a general coordinate system $\left(u^1, u^2, \ldots, u^N\right)$ in a general N-space are related to N other quantities $\overline{C}^1, \overline{C}^2, \ldots, \overline{C}^N$ of another general coordinate system $\left(\overline{u}^1, \overline{u}^2, \ldots, \overline{u}^N\right)$ by the above transformation equations, then the quantities are called *components of a contravariant vector* or a *contravariant tensor of rank one.*

11. Proof of the result in section 9

Recall that

$$\vec{A} = C_1 \nabla u^1 + C_2 \nabla u^2 + C_3 \nabla u^3$$

$$\vec{A} = \bar{C}_1 \nabla \bar{u}^1 + \bar{C}_2 \nabla \bar{u}^2 + \bar{C}_3 \nabla \bar{u}^3$$

We have

$$\bar{u}^1 = \bar{u}^1(u^1, u^2, u^3)$$

$$\bar{u}^2 = \bar{u}^2(u^1, u^2, u^3)$$

$$\bar{u}^3 = \bar{u}^3(u^1, u^2, u^3)$$

Then from calculus (using the chain rule)

$$\frac{\partial \bar{u}^1}{\partial x} = \frac{\partial \bar{u}^1}{\partial u^1}\frac{\partial u^1}{\partial x} + \frac{\partial \bar{u}^1}{\partial u^2}\frac{\partial u^2}{\partial x} + \frac{\partial \bar{u}^1}{\partial u^3}\frac{\partial u^3}{\partial x}$$
$$\frac{\partial \bar{u}^1}{\partial y} = \frac{\partial \bar{u}^1}{\partial u^1}\frac{\partial u^1}{\partial y} + \frac{\partial \bar{u}^1}{\partial u^2}\frac{\partial u^2}{\partial y} + \frac{\partial \bar{u}^1}{\partial u^3}\frac{\partial u^3}{\partial y}$$
$$\frac{\partial \bar{u}^1}{\partial z} = \frac{\partial \bar{u}^1}{\partial u^1}\frac{\partial u^1}{\partial z} + \frac{\partial \bar{u}^1}{\partial u^2}\frac{\partial u^2}{\partial z} + \frac{\partial \bar{u}^1}{\partial u^3}\frac{\partial u^3}{\partial z}$$

$$\frac{\partial \bar{u}^2}{\partial x} = \frac{\partial \bar{u}^2}{\partial u^1}\frac{\partial u^1}{\partial x} + \frac{\partial \bar{u}^2}{\partial u^2}\frac{\partial u^2}{\partial x} + \frac{\partial \bar{u}^2}{\partial u^3}\frac{\partial u^3}{\partial x}$$
$$\frac{\partial \bar{u}^2}{\partial y} = \frac{\partial \bar{u}^2}{\partial u^1}\frac{\partial u^1}{\partial y} + \frac{\partial \bar{u}^2}{\partial u^2}\frac{\partial u^2}{\partial y} + \frac{\partial \bar{u}^2}{\partial u^3}\frac{\partial u^3}{\partial y}$$
$$\frac{\partial \bar{u}^2}{\partial z} = \frac{\partial \bar{u}^2}{\partial u^1}\frac{\partial u^1}{\partial z} + \frac{\partial \bar{u}^2}{\partial u^2}\frac{\partial u^2}{\partial z} + \frac{\partial \bar{u}^2}{\partial u^3}\frac{\partial u^3}{\partial z}$$

$$\frac{\partial \bar{u}^3}{\partial x} = \frac{\partial \bar{u}^3}{\partial u^1}\frac{\partial u^1}{\partial x} + \frac{\partial \bar{u}^3}{\partial u^2}\frac{\partial u^2}{\partial x} + \frac{\partial \bar{u}^3}{\partial u^3}\frac{\partial u^3}{\partial x}$$
$$\frac{\partial \bar{u}^3}{\partial y} = \frac{\partial \bar{u}^3}{\partial u^1}\frac{\partial u^1}{\partial y} + \frac{\partial \bar{u}^3}{\partial u^2}\frac{\partial u^2}{\partial y} + \frac{\partial \bar{u}^3}{\partial u^3}\frac{\partial u^3}{\partial y}$$
$$\frac{\partial \bar{u}^3}{\partial z} = \frac{\partial \bar{u}^3}{\partial u^1}\frac{\partial u^1}{\partial z} + \frac{\partial \bar{u}^3}{\partial u^2}\frac{\partial u^2}{\partial z} + \frac{\partial \bar{u}^3}{\partial u^3}\frac{\partial u^3}{\partial z}$$

Recall that

$$\nabla \bar{u}^1(x,y,z) = \frac{\partial \bar{u}^1}{\partial x}\vec{i} + \frac{\partial \bar{u}^1}{\partial y}\vec{j} + \frac{\partial \bar{u}^1}{\partial z}\vec{k}$$

$$\nabla \bar{u}^2(x,y,z) = \frac{\partial \bar{u}^2}{\partial x}\vec{i} + \frac{\partial \bar{u}^2}{\partial y}\vec{j} + \frac{\partial \bar{u}^2}{\partial z}\vec{k}$$

$$\nabla \bar{u}^3(x,y,z) = \frac{\partial \bar{u}^3}{\partial x}\vec{i} + \frac{\partial \bar{u}^3}{\partial y}\vec{j} + \frac{\partial \bar{u}^3}{\partial z}\vec{k}$$

$$\nabla u^1(x,y,z) = \frac{\partial u^1}{\partial x}\vec{i} + \frac{\partial u^1}{\partial y}\vec{j} + \frac{\partial u^1}{\partial z}\vec{k}$$

$$\nabla u^2(x,y,z) = \frac{\partial u^2}{\partial x}\vec{i} + \frac{\partial u^2}{\partial y}\vec{j} + \frac{\partial u^2}{\partial z}\vec{k}$$

$$\nabla u^3(x,y,z) = \frac{\partial u^3}{\partial x}\vec{i} + \frac{\partial u^3}{\partial y}\vec{j} + \frac{\partial u^3}{\partial z}\vec{k}$$

Substituting and rearranging, we find

$$C_1 \nabla u^1 + C_2 \nabla u^2 + C_3 \nabla u^3 =$$

$$(C_1 \frac{\partial u^1}{\partial x} + C_2 \frac{\partial u^2}{\partial x} + C_3 \frac{\partial u^3}{\partial x})\vec{i} +$$

$$(C_1 \frac{\partial u^1}{\partial y} + C_2 \frac{\partial u^2}{\partial y} + C_3 \frac{\partial u^3}{\partial y})\vec{j} +$$

$$(C_1 \frac{\partial u^1}{\partial z} + C_2 \frac{\partial u^2}{\partial z} + C_3 \frac{\partial u^3}{\partial z})\vec{k}$$

$$\bar{C}_1\ \nabla\bar{u}^1 + \bar{C}_2\ \nabla\bar{u}^2 + \bar{C}_3\ \nabla\bar{u}^3 =$$

$$(\bar{C}_1 \frac{\partial\bar{u}^1}{\partial x} + \bar{C}_2 \frac{\partial\bar{u}^2}{\partial x} + \bar{C}_3 \frac{\partial\bar{u}^3}{\partial x})\vec{i} +$$

$$(\bar{C}_1 \frac{\partial\bar{u}^1}{\partial y} + \bar{C}_2 \frac{\partial\bar{u}^2}{\partial y} + \bar{C}_3 \frac{\partial\bar{u}^3}{\partial y})\vec{j} +$$

$$(\bar{C}_1 \frac{\partial\bar{u}^1}{\partial z} + \bar{C}_2 \frac{\partial\bar{u}^2}{\partial z} + \bar{C}_3 \frac{\partial\bar{u}^3}{\partial z})\vec{k}$$

Equating coefficients, we find

$$C_1 \frac{\partial u^1}{\partial x} + C_2 \frac{\partial u^2}{\partial x} + C_3 \frac{\partial u^3}{\partial x} = \bar{C}_1 \frac{\partial\bar{u}^1}{\partial x} + \bar{C}_2 \frac{\partial\bar{u}^2}{\partial x} + \bar{C}_3 \frac{\partial\bar{u}^3}{\partial x}$$

$$C_1 \frac{\partial u^1}{\partial y} + C_2 \frac{\partial u^2}{\partial y} + C_3 \frac{\partial u^3}{\partial y} = \bar{C}_1 \frac{\partial\bar{u}^1}{\partial y} + \bar{C}_2 \frac{\partial\bar{u}^2}{\partial y} + \bar{C}_3 \frac{\partial\bar{u}^3}{\partial y}$$

$$C_1 \frac{\partial u^1}{\partial z} + C_2 \frac{\partial u^2}{\partial z} + C_3 \frac{\partial u^3}{\partial z} = \bar{C}_1 \frac{\partial\bar{u}^1}{\partial z} + \bar{C}_2 \frac{\partial\bar{u}^2}{\partial z} + \bar{C}_3 \frac{\partial\bar{u}^3}{\partial z}$$

Substituting the set of nine equations above in the set of three equations above, rearranging, and equating coefficients of

$$\frac{\partial u^1}{\partial x}, \frac{\partial u^2}{\partial x}, \frac{\partial u^3}{\partial x}, \frac{\partial u^1}{\partial y}, \frac{\partial u^2}{\partial y}, \frac{\partial u^3}{\partial y}, \frac{\partial u^1}{\partial z}, \frac{\partial u^2}{\partial z}, \frac{\partial u^3}{\partial z}$$

on each side, we find

$$C_1 = \bar{C}_1 \frac{\partial \bar{u}^1}{\partial u^1} + \bar{C}_2 \frac{\partial \bar{u}^2}{\partial u^1} + \bar{C}_3 \frac{\partial \bar{u}^3}{\partial u^1}$$

$$C_2 = \bar{C}_1 \frac{\partial \bar{u}^1}{\partial u^2} + \bar{C}_2 \frac{\partial \bar{u}^2}{\partial u^2} + \bar{C}_3 \frac{\partial \bar{u}^3}{\partial u^2}$$

$$C_3 = \bar{C}_1 \frac{\partial \bar{u}^1}{\partial u^3} + \bar{C}_2 \frac{\partial \bar{u}^2}{\partial u^3} + \bar{C}_3 \frac{\partial \bar{u}^3}{\partial u^3}$$

or

$$C_p = \frac{\partial \bar{u}^q}{\partial u^p} \bar{C}_q$$

or, by interchanging the coordinates, we can show that

$$\bar{C}_p = \frac{\partial u^q}{\partial \bar{u}^p} C_q$$

The above results lead us to adopt the following definition: If N quantities $C_1, C_2, ..., C_N$ of a general coordinate system $(u^1, u^2, ..., u^N)$ in a general N-space are related to N other quantities $\bar{C}_1, \bar{C}_2, ..., \bar{C}_N$ of another general coordinate system $(\bar{u}^1, \bar{u}^2, ..., \bar{u}^N)$ by the above transformation equations, then the quantities are called *components of a covariant vector* or a *covariant tensor of rank one.*

12. General differentiable *N*-spaces and general coordinates. Curves and surfaces

In this book we will use the word *space* even if other books often use other words, such as *manifold* or *hyperspace*. Tensor calculus works in a general *differentiable* space of *N* dimensions, where $N \geq 2$. Note that tensor calculus can work in a *non-metrical space*, in which the concept of *distance* never enters. It is a remarkable feature of tensor calculus that no essential simplification is obtained by taking a small value of *N*.

Let $(u^1, u^2, \dots, u^N)$ and $(\overline{u}^1, \overline{u}^2, \dots, \overline{u}^N)$ be coordinates of a point in a general differential *N*-space in two different coordinate systems. Suppose there exists *N* independent equations between the coordinates of the two coordinate systems having the form

$$\overline{u}^k = \overline{u}^k(u^1, u^2, \dots, u^N)$$

where *k* = 1, 2, ... , *N*. Of course, it is supposed that the functions above are single valued, continuous, and have continuous derivatives. Then conversely we have

$$u^k = u^k(\overline{u}^1, \overline{u}^2, \dots, \overline{u}^N)$$

where *k* = 1, 2, ... , *N*. Of course, we suppose that these equations too are single-valued, continuous, and have continuous derivatives. The two sets of equations above define a *transformation of coordinates* from one coordinate system to another.

Given a differentiable *N*-space, we define a *curve* by the parametric equations

$$u^a = u^a(t)$$

where *a* = 1, 2,..., *N*, and t is the parameter. Similarly, a *surface* (*subspace*) of *M* dimensions $(M \langle N)$ is given by the parametric equations

$$u^a = u^a(t^1, t^2, \dots, t^M)$$

If $M = N - 1$, the surface is called a *hypersurface.* In this case the parameters can be eliminated from the N equations to give one equation connecting the coordinates, i.e.

$$F(u^1, u^2, \ldots, u^N) = 0$$

Thus, if a point is restricted to lie in this hypersurface ((N – 1)-*subspace*), then its coordinates must satisfy one *constraint,* namely, $F(u^1, u^2, \ldots, u^N) = 0$. Of course, this equation is an alternative to the parametric equations.

13. Contravariant tensors of rank one

Consider, in a differentiable *N*-space, a point *P* with coordinates u^a, where a = 1, 2, ..., N, and a neighboring point *Q* with coordinates $u^a + du^a$. These two points define an *infinitesimal displacement* $\overrightarrow{PQ}$ or *infinitesimal vector* $\overrightarrow{PQ}$. For the u^a coordinate system $\overrightarrow{PQ}$ is described by the quantities du^a, which are called the *components* of $\overrightarrow{PQ}$ in the u^a coordinate system. *The vector* $\overrightarrow{PQ}$ *is not to be regarded as free, but as being associated with (or attached to) the point P, and it is to be considered as having an absolute meaning, but the numbers which describe it depend on the coordinate system employed.* Let $\bar{u}^a$ be the coordinates of *P* in another coordinate system: the $\bar{u}^a$ coordinate system. Then from calculus we have (calculated at P)

$$d\bar{u}^a = \frac{\partial \bar{u}^a}{\partial u^1} du^1 + \frac{\partial \bar{u}^a}{\partial u^2} du^2 + \ldots + \frac{\partial \bar{u}^a}{\partial u^N} du^N$$

or

$$d\bar{u}^a = \frac{\partial \bar{u}^a}{\partial u^b} du^b$$

Of course, we also have

$$du^a = \frac{\partial u^a}{\partial \bar{u}^b} d\bar{u}^b$$

Clearly, from section 10 above, the infinitesimal displacement or infinitesimal vector du^a attached to the point *P* with coordinates u^a is a *contravariant tensor of rank one.* Note that the location of the index a (or b) is appropriate. A tensor is a geometrical quantity defined in terms of its transformation properties under a coordinate transformation. When we make a statement about tensors we wish it to hold for *all* coordinate systems.

If N quantities A^a (where a = 1, 2, … , N) in a coordinate system u^a are related to N other quantities $\overline{A}^a$ in another coordinate system $\overline{u}^a$ by the transformation equations

$$\overline{A}^a = \frac{\partial \overline{u}^a}{\partial u^b} A^b$$

and

$$A^a = \frac{\partial u^a}{\partial \overline{u}^b} \overline{A}^b$$

they are called components of a *contravariant vector* or *contravariant tensor of rank one.* Instead of speaking of "a tensor whose components are A^a" we shall often refer to "the tensor A^a." No confusion should arise from this. Note that quantities with bars and quantities without bars can be interchanged.

Examples: The velocity of a fluid at any point has components

$$\frac{dx^a}{dt}$$

(where t is the parameter time) in the coordinate system x^a In the coordinate system $\overline{x}^a$ the velocity is

$$\frac{d\overline{x}^a}{dt}$$

But from calculus (by the chain rule) we have

$$\frac{d\overline{x}^a}{dt} = \frac{\partial \overline{x}^a}{\partial x^b}\frac{dx^b}{dt}$$

The velocity is a contravariant tensor of rank one or a contravariant vector. But note that the quantity

$$\frac{dv^a}{dt} = \frac{d^2x^a}{dt^2}$$

is not a tensor and so cannot represent acceleration in all coordinate systems. However, acceleration represented by

$$a^k = \frac{\delta v^k}{\delta t} \quad \text{(= the intrinsic derivative of } v\text{)}$$

is a contravariant tensor of rank one (see section 47) and so can represent acceleration in all coordinate systems. Again, note that the location of the index a (or b or k) is appropriate.

By the way, the tangent vector $T^p = \frac{dx^p}{ds}$ to curve C whose equations are $x^p = x^p(s)$, where the parameter s is the arc length, is a contravariant tensor of rank one.

14. Covariant tensors of rank one

Let $\phi = \phi\,(u^1, u^2, \ldots, u^N)$ be a continuous and differentiable function on an *N*-space with the u^a coordinate system and the $\bar{u}^a$ coordinate system, and the functions $u^a = u^a\,(\bar{u}^1, \bar{u}^2, \ldots, \bar{u}^N)$ Then we have

$$\phi = \phi\,(u^1\,(\bar{u}^1, \bar{u}^2, \ldots, \bar{u}^N), u^2\,(\bar{u}^1, \bar{u}^2, \ldots, \bar{u}^N), \ldots, u^N\,(\bar{u}^1, \bar{u}^2, \ldots, \bar{u}^N))$$

From calculus, using the function of a function rule, we obtain

$$\frac{\partial \phi}{\partial \bar{u}^a} = \frac{\partial \phi}{\partial u^b}\frac{\partial u^b}{\partial \bar{u}^a}$$

or

$$\frac{\partial \phi}{\partial \bar{u}^a} = \frac{\partial u^b}{\partial \bar{u}^a}\frac{\partial \phi}{\partial u^b}$$

is a covariant tensor of rank one.

If N quantities A_a (where a = 1, 2,…, N) in a coordinate system u^a are related to N other quantities $\bar{A}_a$ in another coordinate system $\bar{u}^a$ by the transformation equations

$$\bar{A}_a = \frac{\partial u^b}{\partial \bar{u}^a} A_b$$

and

$$A_a = \frac{\partial \bar{u}^b}{\partial u^a} \bar{A}_b$$

they are called components of a *covariant vector* or *covariant tensor of rank one.* Instead of speaking of "a tensor whose components are A_a" we shall often refer to "the tensor A_a".

No confusion should arise from this. Note that quantities with bars and quantities without bars can be interchanged.

Examples: If we let the mass M of a particle be an invariant, then the (covariant) force $F_k = Ma_k$ is a covariant tensor of rank one, because a tensor when multiplied by an invariant is again a tensor of the same type (see section 23).

Consider the differential expression $dW = F_n\, dx^n$. Here $F_n\, dx^n$ is the *inner product* of the two tensors F_k and dx^a (se section 26). This product is an invariant (see section 18).

15. Mixed tensors of the second rank

The following *product* of two tensors involves ordinary multiplication of the components of the tensor and is called the *outer product*:

$$\bar{A}^p \bar{B}_r = \left(\frac{\partial \bar{u}^p}{\partial u^q} A^q \right) \left(\frac{\partial u^s}{\partial \bar{u}^r} B_s \right)$$

$$= \left(\frac{\partial \bar{u}^p}{\partial u^1} A^1 + \frac{\partial \bar{u}^p}{\partial u^2} A^2 +, \dots, + \frac{\partial \bar{u}^p}{\partial u^N} A^N \right) \left(\frac{\partial u^1}{\partial \bar{u}^r} B_1 + \frac{\partial u^2}{\partial \bar{u}^r} B_2 +, \dots, + \frac{\partial u^N}{\partial \bar{u}^r} B_N \right)$$

$$= \frac{\partial \bar{u}^p}{\partial u^q} \frac{\partial u^s}{\partial \bar{u}^r} A^q B_s$$

If we write $A^p B_r = C_r^p$, we have

$$\bar{C}_r^p = \frac{\partial \bar{u}^p}{\partial u^q} \frac{\partial u^s}{\partial \bar{u}^r} C_s^q$$

and

$$C_r^p = \frac{\partial u^p}{\partial \bar{u}^q} \frac{\partial \bar{u}^s}{\partial u^r} \bar{C}_s^q$$

A set of quantities $C_r{}^p$ are said to be the components of a *mixed tensor* of *rank* two if they transform according to the equations above.

Examples: The covariant derivative (see section 46) of a contravariant tensor of rank one is a mixed tensor of rank two. The *Kronecker delta* (see section 34) is a mixed tensor of rank two.

16. Contravariant and covariant tensors of rank two

The outer product (see section 15) of two contravariant tensors of rank one is (we omit the proof)

$$\bar{A}^p \bar{B}^r = \frac{\partial \bar{u}^p}{\partial u^q} \frac{\partial \bar{u}^r}{\partial u^s} A^q B^s$$

or (see section 15)

$$\bar{C}^{pr} = \frac{\partial \bar{u}^p}{\partial u^q} \frac{\partial \bar{u}^r}{\partial u^s} C^{qs}$$

and

$$C^{pr} = \frac{\partial u^p}{\partial \bar{u}^q} \frac{\partial u^r}{\partial \bar{u}^s} \bar{C}^{qs}$$

A set of quantities C^{pr} are said to be the *components of a contravariant tensor of rank two* if they transform according to the equations above. Examples: The *contravariant metric tensor*, denoted by g^{ij} is a contravariant tensor of rank two (see section31). The contravariant *Einsten's tensor*, denoted by R^{ij}, is a contravariant tensor of rank two.

The outer product (see section 15) of two covariant tensors of rank one is (we omit the proof)

$$\bar{A}_p \bar{B}_r = \frac{\partial u^q}{\partial \bar{u}^p} \frac{\partial u^s}{\partial \bar{u}^r} A_q B_s$$

or (see section 23)

$$\bar{C}_{pr} = \frac{\partial u^q}{\partial \bar{u}^p} \frac{\partial u^s}{\partial \bar{u}^r} C_{qs}$$

and

$$C_{pr} = \frac{\partial \bar{u}^q}{\partial u^p} \frac{\partial \bar{u}^s}{\partial u^r} \bar{C}_{qs}$$

A set of quantities C_{pr} are said to be the *components of a covariant tensor of rank two* if they transform according to the equations above. Examples: The *covariant derivative* (see section 46) of a covariant tensor of rank one is a covariant tensor of rank two. The covariant *Ricci tensor*, denoted by R_{ij}, is a covariant tensor of rank two.

17. Tensors of rank greater than two

Tensors of rank greater than two are easily defined. The outer product of two tensors (ordinary multiplication of the components of the tensor) is a tensor whose rank is the sum of the ranks of the given tensors. For example

$$\overline{A}_i^p \overline{B}^r \overline{C}_j^m = \frac{\partial \overline{u}^p}{\partial u^q} \frac{\partial \overline{u}^r}{\partial u^s} \frac{\partial \overline{u}^m}{\partial u^t} \frac{\partial u^k}{\partial \overline{u}^i} \frac{\partial u^l}{\partial \overline{u}^j} A_k^q B^s C_l^t$$

or (see section 15 and 16)

$$\overline{D}_{ij}^{prm} = \frac{\partial \overline{u}^p}{\partial u^q} \frac{\partial \overline{u}^r}{\partial u^s} \frac{\partial \overline{u}^m}{\partial u^t} \frac{\partial u^k}{\partial \overline{u}^i} \frac{\partial u^l}{\partial \overline{u}^j} D_{kl}^{qst}$$

This tensor is a mixed tensor of rank 5, contravariant of order 3 and covariant of order 2. As always, quantities with bars and quantities without bars can be interchanged.

The definitions of tensors of the third, fourth, or higher orders will at once suggest themselves, and it is unnecessary to write them down here.

18. Invariants or scalars (tensors of rank zero)

The definitions of tensors above suggest that there should be a tensor of zero order, a single quantity, transforming according to the transformation equation

$$\bar{A}\left(\bar{u}^1,\bar{u}^2,\ldots,\bar{u}^N\right)=A\left(u^1,u^2,\ldots,u^N\right)$$

Such a quantity is called an *invariant* or a *scalar* with respect to the above coordinate transformation. It's also called a *tensor of rank zero*. Its value is independent of the coordinate system used. The obvious example is a pure number, like Π or zero. Another example is *ds* given by the equation $ds^2 = g_{pq}dx^p dx^q$. See section 31. Note that an invariant is a scalar, but not necessarily a constant.

19. Tensor equations

Suppose, for example, that we are given that a tensor $A_{pq} = 0$. Then it is an immediate consequence of the transformation equations that $\bar{A}_{pq} = 0$. More generally, suppose we are given that, for example, $A_{pq} = B_{pq}$. Then $\bar{A}_{pq} = \bar{B}_{pq}$. This, too, follows directly from the transformation equations. In short, a tensor equation which holds in one particular coordinate system necessarily holds in *all* coordinate systems. This result is the reason why tensors are very important in physics.

20. Tensor fields

A tensor may be given at a single point of an *N*-space, or it may be given along a curve or throughout a subspace of the *N*-space, or, of course, throughout the *N*-space itself. In the last three cases we say that a *tensor field* has been defined. The distinction between a tensor field and a tensor is not always made completely clear. We often refer to tensors fields as tensors. A tensor field $T_{b...}^{a...}(u^1, u^2, ..., u^N)$ is called continuous or differentiable if $T_{b...}^{a...}(u^1, u^2, ..., u^N)$ in all coordinate systems are continuous or differentiable functions of the coordinates. A tensor field $T_{b...}^{a...}(u^1, u^2, ..., u^N)$ is called *smooth* if $T_{b...}^{a...}(u^1, u^2, ..., u^N)$ in all coordinate systems are differentiable to *all* orders. Note that a tensor (or a tensor field) is not just the set of its components in one special coordinate system but all the possible sets under any transformation of coordinates.

21. Addition of tensors

It follows directly from the transformation equations that the *sum* of two or more tensors of the same rank and type is also a tensor of the same rank and type. Example: If A^{mp}_{q} and B^{mp}_{q} are tensors, then $C^{mp}_{q} = A^{mp}_{q} + B^{mp}_{q}$ is also a tensor. Note that in adding tensors we use only tensors of a single type, and add components with the same literal suffixes. Of course, these restrictive rules on addition are introduced in order that the results of addition may themselves be tensors.

22. Subtraction of tensors

The *difference* of two tensors of the same rank and type is also a tensor of the same rank and type. Example: It follows directly from the transformation equations that if A_q^{mp} and B_q^{mp} are tensors, then $D_q^{mp} = A_q^{mp} - B_q^{mp}$ is also a tensor. Note that in subtracting tensors we use only tensors of a single type, and subtract components with the same literal suffixes. Of course, these restrictive rules on subtraction are introduced in order that the results of subtraction may themselves be tensors.

23. Outer Multiplication of tensors

In adding or subtracting tensors we use only tensors of a single type, and add or subtract components with the same literal suffixes. This is not the case in *outer multiplication.* Outer multiplication involves ordinary multiplication of the components. Example: $C^{prm}_{qs} = A^{pr}_{q} B^{m}_{s}$ is the *outer product* of A^{pr}_{q} and B^{m}_{s}. The only restriction here is that we never multiply two components with the same literal suffix at the same level in each. Of course, this restrictive rule on outer multiplication is introduced in order that the results of outer multiplication may themselves be tensors. It follows directly from the transformation equations that the outer product is a tensor of the type indicated.

A tensor when multiplied by an invariant is again a tensor of the same rank and type. Example: $\phi A^{a}_{bc} = B^{a}_{bc}$. And an invariant when multiplied by an invariant is again an invariant. Example: $\phi_1\phi_2 = \phi_3$. Of course, these results, too, follow directly from the transformation equations.

24. Division of tensors

Not every tensor can be written as a product of two tensors of lower rank. Thus, *division* of tensors is not always possible. Of course, division of tensors follows directly from multiplication of tensors.

25. Contraction of tensors

If one contravariant and one covariant index of a tensor are set equal, the result indicates that a summation over the equal indices is to be taken. (Recall the summation convention.) It follows from the transformation equations that the resulting sum is a tensor of rank two less than that of the original tensor. Examples: If the original tensor is A^{mpr}_{qs} and we set $r = s$, we obtain $A^{mpr}_{qr} = B^{mp}_{q}$. By a second contraction, setting $p = q$, we obtain $B^{mp}_{p} = C^{m}$. Note: Contraction cannot be applied to suffixes at the same level.

26. Inner multiplication of tensors

By the operation of outer multiplication of two tensors followed by a contraction we obtain a new tensor called an *inner product* of the given tensors. (The tensor character of the result follows from the transformation equations.) This process is called *inner multiplication* Generally, inner multiplication consists in a combination of outer multiplication with one or more contractions. Examples: The outer product of the tensors A_q^{mp} and B_{st}^r is $A_q^{mp} B_{st}^r$. If we let $q = r$ (one contraction), we obtain the inner product $A_q^{mp} B_{st}^q = C_{st}^{mp}$. This operation may be characterized as a mixed one, being "outer" with respect to the indices m, p, s, and t, and "inner" with respect to the indices q and r. If we let $q = r$ and $p = s$ (double contraction), we obtain the inner product $A_q^{mp} B_{pt}^q = C_t^m$. (The tensor character of the result follows from the transformation equations.) This operation may be characterized as being "outer" with respect to the indices m and t, and "inner" with respect to the indices p, q, r, and s. If we let $q = r$, $p=s$, and $m= t$ (triple contraction), we obtain the inner product $A_q^{mp} B_{pm}^q = C_m^m = C$, a tensor of rank zero (an invariant). Again, of course, the tensor character of the result follows from the transformation equations. Note that it is customary to use the same symbol (here C) for the contracted tensor as for the original tensor.

Since $A^p B_p$ (the inner product of the tensors (vectors) A^p and B_q) is a scalar (an invariant), it is often called the *scalar product* of the tensors (vectors) A^p and B_q.

27. Quotient law of tensors

Suppose that we do not know whether a quantity X is a tensor or not. If an inner product of X with an arbitrary tensor is a tensor, then X is a tensor. This theorem is called the *quotient law*. It follows from the transformation equations.

28. Symmetry

We say that a tensor is *symmetric* with respect to a pair of suffixes (both superscripts or both subscripts) if the value of the component is unchanged on interchanging these suffixes. The property of *symmetry* is conserved under transformation of coordinates. Example: If $A_{rs} = A_{sr}$ is a symmetric tensor in a given coordinate system, then the difference $A_{rs} - A_{sr}$ is itself a tensor. (Note that in subtracting tensors we use only tensors of a single type, and subtract components with the same literal suffixes, although *these need not occur in the same order.*) But this tensor (the difference) vanish (= 0) in the given coordinate system, and so vanishes in all coordinate systems.

29. Skew-symmetry (or antisymmetry)

We say that a tensor is *skew-symmetric* or *antisymmetric* with respect to a pair of suffixes (both superscripts or both subscripts) if the component changes sign without changing its absolute value on interchanging these suffixes. The property of *skew-symmetry* is conserved under transformation of coordinates. Example: If $A_{rs} = -A_{sr}$ is a skew-symmetric tensor in a given coordinate system, then the sum $A_{rs} + A_{sr}$ is itself a tensor. (Note that in subtracting tensors we use only tensors of a single type, and subtract components with the same literal suffixes, although *these need not occur in the same order.*) But this tensor (the difference) vanish (= 0) in the given coordinate system, and so vanishes in all coordinate systems.

30. How to perform calculations

It is a remarkable feature of tensor calculus that the general aspects of a space of, say, three million dimensions is as easy to discuss as the general aspects of a space of, say, three dimensions. It can be proved, although we shall not do it, that when we are treating expressions where the summation and range conventions are employed, we can, in general perform the mathematical operations just as in elementary algebra ("forgetting" the conventions).

The fundamental operations with tensors: addition, subtraction, outer and inner multiplication, and division, satisfy, in general, the well known axioms of elementary algebra. Examples:

$$1A^{a\ldots}_{b\ldots} = A^{a\ldots}_{b\ldots}$$

$$A^{a\ldots}_{b\ldots} + 0 = A^{a\ldots}_{b\ldots}$$

$$A^{a\ldots}_{b\ldots} + \left(-A^{a\ldots}_{b\ldots}\right) = 0$$

$$A^{mp}_{q} + B^{mp}_{q} = B^{mp}_{q} + A^{mp}_{q}$$

$$A^{mp}_{p} + B^{mp}_{p} = B^{mp}_{p} + A^{mp}_{p}$$

$$A^{mp}_{q} + \left(B^{mp}_{q} + C^{mp}_{q}\right) = \left(A^{mp}_{q} + B^{mp}_{q}\right) + C^{mp}_{q}$$

$$A^{mp}_{p} + \left(B^{mp}_{p} + C^{mp}_{p}\right) = \left(A^{mp}_{p} + B^{mp}_{p}\right) + C^{mp}_{p}$$

$$A^{p}_{q} B_{rs} = B_{rs} A^{p}_{q} = C^{p}_{rsq}$$

$$A^{p} B_{p} = B_{p} A^{p} = C^{p}_{p} = C$$

$$A^{mp}_{q}\left(B^{rs} C_{t}\right) = \left(A^{mp}_{q} B^{rs}\right) C_{t} = D^{mprs}_{qt}$$

$$A^{b}_{a}\left(B^{c}_{b} C^{d}_{c}\right) = \left(A^{b}_{a} B^{c}_{b}\right) C^{d}_{c} = D^{d}_{a}$$

$$A^{rs}\left(B^{mp}_{q} + C^{mp}_{q}\right) = A^{rs} B^{mp}_{q} + A^{rs} C^{mp}_{q}$$

$$A^{qr}\left(B^{mp}_{q} + C^{mp}_{q}\right) = A^{qr} B^{mp}_{q} + A^{qr} C^{mp}_{q}$$

Tensors do not satisfy the existence of reciprocals-axiom, but if, for example,

$$A^r_{pq} B^{qs}_r = 1$$

where A^r_{pq} is an arbitrary tensor, then B^{qs}_r is a tensor.

The axioms are tensor equations. Therefore, if they are true for one special coordinate system, then they are true for all coordinate systems. And the theorems follow logically from the axioms, and the theorems are tensor equations, and so hold in all coordinate systems. Therefore, tensor calculus satisfy, in general, the basic axioms and theorems of mathematics. An example: If X(p, q, r) is a quantity such that X(p, q, r) $B^{qn}_r = 0$ for an arbitrary tensor B^{qn}_r, then X(p, q, r) = 0 identically.

Note that scalars, vectors, and tensors are invariant under coordinate transformations, but vector components and tensor components are not; this meaning that the scalars, vectors, and tensors are geometric quantities which have a coordinate independent meaning.

As we have noted several times already: Quantities with bars and quantities without bars can be interchanged. We shall not give the general proof for this result. But to illustrate, we shall prove a special case of the result:

If $\bar{C}^p = \dfrac{\partial \bar{x}^p}{\partial x^q} C^q$ then $C^q = \dfrac{\partial x^q}{\partial \bar{x}^p} r\bar{C}^p$ or $C^p = \dfrac{\partial x^p}{\partial \bar{x}^q} \bar{C}^q$

Proof:

We have $\bar{C}^p = \dfrac{\partial \bar{x}^p}{\partial x^q} C^q$

Then (see section 34)

$$\frac{\partial x^r}{\partial \bar{x}^p} \bar{C}^p = \frac{\partial x^r}{\partial \bar{x}^p} \frac{\partial \bar{x}^p}{\partial x^q} C^q = \delta^r_q C^q = C^r$$

Putting $r = q$ we have

$$\frac{\partial x^q}{\partial \bar{x}^p} \bar{C}^p = C^q \quad \text{or} \quad C^q = \frac{\partial x^q}{\partial \bar{x}^p} \bar{C}^p \quad \text{or} \quad C^p = \frac{\partial x^p}{\partial \bar{x}^q} \bar{C}^q$$

Transitivity of tensor character: In general differentiable *N*-space, let there be three coordinate systems: $x^r, \overline{x}^r, \overline{\overline{x}}^r$. Given that

$$\overline{A}_{ab} = \frac{\partial x^c}{\partial \overline{x}^a} \frac{\partial x^d}{\partial \overline{x}^b} A_{cd}$$

$$\text{and} \quad \overline{\overline{A}}_{ab} = \frac{\partial \overline{x}^c}{\partial \overline{\overline{x}}^a} \frac{\partial \overline{x}^d}{\partial \overline{\overline{x}}^b} \overline{A}_{cd}$$

$$\text{then} \quad \overline{\overline{A}}_{ab} = \frac{\partial x^c}{\partial \overline{\overline{x}}^a} \frac{\partial x^d}{\partial \overline{\overline{x}}^b} A_{cd}$$

This follows from the formulae of tensor transformation, and is easy to show. The transitivity of tensor character holds quite generally.

31. The line element and the metric tensor. Riemannian space

From calculus: In rectangular coordinates (x, y, z) in Euclidean 3-space the differential of arc length ds (or the element of arc length ds) is obtained from

$$ds^2 = dx^2 + dy^2 + dz^2$$

By transforming to general curvilinear coordinates (u^1, u^2, u^3) we have

$$ds^2 = g_{pg} du^p du^q$$

where $g_{pq} = \dfrac{\partial \vec{r}}{\partial u^p} \bullet \dfrac{\partial \vec{r}}{\partial u^q}$ and, of course, $p = 1, 2, 3$ and $q = 1, 2, 3$.

Proof

From calculus

$$d\vec{r} = \frac{\partial \vec{r}}{\partial u^1} du^1 + \frac{\partial \vec{r}}{\partial u^2} du^2 + \frac{\partial \vec{r}}{\partial u^3} du^3$$

Then (from calculus)

$$ds^2 = d\vec{r} \square d\vec{r}$$

$$= \frac{\partial \vec{r}}{\partial u^1} \bullet \frac{\partial \vec{r}}{\partial u^1} \left(du^1\right)^2 + \frac{\partial \vec{r}}{\partial u^1} \bullet \frac{\partial \vec{r}}{\partial u2^2} du^1 du^2 + \frac{\partial \vec{r}}{\partial u^1} \bullet \frac{\partial \vec{r}}{\partial u^3} du^1 du^3$$

$$+ \frac{\partial \vec{r}}{\partial u^2} \bullet \frac{\partial \vec{r}}{\partial u^1} du^2 du^1 + \frac{\partial \vec{r}}{\partial u^2} \bullet \frac{\partial \vec{r}}{\partial u^2} \left(du^2\right)^2 + \frac{\partial \vec{r}}{\partial u^2} \bullet \frac{\partial \vec{r}}{\partial u^3} du^2 du^3$$

$$+ \frac{\partial \vec{r}}{\partial u^3} \bullet \frac{\partial \vec{r}}{\partial u^1} du^3 du^1 + \frac{\partial \vec{r}}{\partial u^3} \bullet \frac{\partial \vec{r}}{\partial u^2} du^3 du^2 + \frac{\partial \vec{r}}{\partial u^3} \bullet \frac{\partial \vec{r}}{\partial u^3} \left(du^3\right)^2$$

$$= g_{pq} du^p du^q$$

where

$$g_{pq} = \frac{\partial \vec{r}}{\partial u^p} \cdot \frac{\partial \vec{r}}{\partial u^q} = \frac{\partial x}{\partial u^p}\frac{\partial x}{\partial u^q} + \frac{\partial y}{\partial u^p}\frac{\partial y}{\partial u^q} + \frac{\partial z}{\partial u^p}\frac{\partial z}{\partial u^q}$$

$ds^2 = g_{pg} du^p du^q$ is called the *fundamental quadratic form* or *metric form*. This equation is also known as the *line element*. The quantities g_{pq} are called *metric coefficients*. Observe that $g_{pq} = g_{qp}$ If $g_{pq} = 0$, $p \neq q$, then the curvilinear coordinate system is orthogonal. For orthogonal curvilinear systems (in Euclidean 3-space)

$$ds^2 = g_{11}\left(du^1\right)^2 + g_{22}\left(du^2\right)^2 + g_{33}\left(du^3\right)^2$$

For non-orthogonal or general curvilinear systems (in Euclidean 3-space)

$$\begin{aligned} ds^2 = \\ & g_{11}\left(du^1\right)^2 + g_{12} du^1 du^2 + g_{13} du^1 du^3 \\ & + g_{21} du^2 du^1 + g_{22}\left(du^2\right)^2 + g_{23} du^2 du^3 \\ & + g_{31} du^3 du^1 + g_{32} du^3 du^2 + g_{33}\left(du^3\right)^2 \end{aligned}$$

We know that the outer product $du^p du^q$ is an arbitrary contravariant tensor of rank 2. We know that ds^2 is a tensor of rank zero. We know that $g_{pq} du^p du^q$ is the inner product of g_{pq} and $du^p du^q$. Then, by the quotient law, g_{pq} is a covariant tensor of rank 2. It's called the *metric tensor* or the *fundamental tensor*. Of course, the g_{pq} are functions of the coordinates. Note that the metric tensor g_{pq} is defined to transform in just such a way as to keep ds^2 invariant. Now generalization to general Riemannian *N*-space with general coordinates $(u^1, u^2, \ldots, u^N)$ is immediate. (See below.) We can and always will choose the metric tensor to be symmetric. An *N*-space endowed with a symmetric metric tensor is called *Riemannian*. This definition does not introduce the word "distance". Note that Riemannian geometry is not built up on the concept of finite distance but on the concept of *ds*. In the special case where there exists a transformation of coordinates from u^j to $\overline{u}^k$ such that the metric form is transformed into $ds^2 = \left(d\overline{u}^1\right)^2 + \left(d\overline{u}^2\right)^2 + \ldots + \left(d\overline{u}^N\right)^2 = d\overline{u}^k d\overline{u}^k$, then the space is called *Euclidean N-space.* It is proved that *on every differentiable space (manifold) there exists a Riemannian metric.* On a Riemannian space we can define the length and the energy integral and other geometrical and physical quantities.

As we already know, in a flat Riemannian 3-space we have

$$g_{pq} = \frac{\partial x^1}{\partial \overline{x}^p}\frac{\partial x^1}{\partial \overline{x}^q} + \frac{\partial x^2}{\partial \overline{x}^p}\frac{\partial x^2}{\partial \overline{x}^q} + \frac{\partial x^3}{\partial \overline{x}^p}\frac{\partial x^3}{\partial \overline{x}^q} \qquad p,q = 1,2,3$$

where $x^1 = x, \quad x^2 = y, \quad x^3 = z$ are rectangular coordinates and $\overline{x}^1, \overline{x}^2, \overline{x}^3$ are curvilinear coordinates.

To avoid confusion, we must be careful to frame our definition of g_{pq} in Euclidean *N*-space, with Cartesian coordinates x^r and curvilinear coordinates $\overline{x}^r$, so that this definition agree with the familiar one above:

$$g_{pq} = \frac{\partial x^1}{\partial \overline{x}^p}\frac{\partial x^1}{\partial \overline{x}^q} + \frac{\partial x^2}{\partial \overline{x}^p}\frac{\partial x^2}{\partial \overline{x}^q} + \ldots + \frac{\partial x^N}{\partial \overline{x}^p}\frac{\partial x^N}{\partial \overline{x}^q} = \frac{\partial x^k}{\partial \overline{x}^p}\frac{\partial x^k}{\partial \overline{x}^q}$$

where $p,q = 1,2,\ldots,N$

(See more on g_{pq} in section 36.)

Not every change in the components g_{pq} is associated with a change in the metric of the space. The components g_{pq} also change under a simple transformation of coordinates connected merely with a shift from one system to another in one and the same space.

It should be understood that a Riemannian *N*-space is a Euclidean *N*-space (a flat Riemannian *N*-space) only if it is possible to transform $ds^2 = g_{pq} dx^p dx^q$

into $ds^2 = d\bar{x}^k d\bar{x}^k$ *throughout* the space. Note that, in a flat Riemannian *N*-space, any system of coordinates for which $g_{pq} = \delta^p_q$ and $ds^2 = dx^k dx^k$ is called *homogeneous.* (In ordinary Euclidean 3-space rectangular Cartesian coordinates are homogeneous coordinates.) As long as we keep to homogeneous coordinates, the transformations involved are *orthogonal transformations.* For tensors in Euclidean *N*-space with orthogonal coordinates there is no distinction between contravariant and covariant components. Here, for example, $A^p = A_p$, and $g_{pq} = g^{pq}$.

The tensor field g_{pq} defines a *metric.* A metric can be used to define distances, lengths of vectors, etc. The equation $ds^2 = g_{pq} dx^p dx^q$ is often called the *line element* and g_{pq} is often called the *metric form* or *first fundamental form.* The square of the *length* or *norm* of a contravariant vector X^p is defined by $X^2 = g_{pq} X^p X^q$ (see section 41). The metric is said to be *positive definite* if, for all vectors X^p, $X^2 \succ 0$. The metric is said to be *negative definite* if, for all vectors X^p, $X^2 \prec 0$. Otherwise, the metric is called *indefinite.* Note that *ds* is the length (norm) of the infinitesimal vector dx^r, and that $(ds)^2$ is conventionally written ds^2. We assume that $ds \succ 0$. If $ds = 0$ the differential equations of a geodesic (as discussed in sections 47 and 51) become meaningless. For the sake of simplicity we shall also assume that the coordinates, the components g_{pq}, and *ds* are real numbers.

32. The Jacobian of coordinate transformations in Euclidean 3-space

Let (x, y, z) be rectangular Cartesian coordinates and let (u^1, u^2, u^3) be orthogonal curvilinear coordinates in Euclidean 3-space. Let both coordinate systems be right-handed. Then we have that the *Jacobian* (J) of x, y, z with respect to u^1, u^2, u^3 is

$$J\left(\frac{x, y, z}{u^1, u^2, u^3}\right) = \begin{vmatrix} \frac{\partial x}{\partial u^1} & \frac{\partial y}{\partial u^1} & \frac{\partial z}{\partial u^1} \\ \frac{\partial x}{\partial u^2} & \frac{\partial y}{\partial u^2} & \frac{\partial z}{\partial u^2} \\ \frac{\partial x}{\partial u^3} & \frac{\partial y}{\partial u^3} & \frac{\partial z}{\partial u^3} \end{vmatrix}$$

$$= \left(\frac{\partial x}{\partial u^1}\vec{i} + \frac{\partial y}{\partial u^1}\vec{j} + \frac{\partial z}{\partial u^1}\vec{k}\right) \cdot \left[\left(\frac{\partial x}{\partial u^2}\vec{i} + \frac{\partial y}{\partial u^2}\vec{j} + \frac{\partial z}{\partial u^2}\vec{k}\right) \times \left(\frac{\partial x}{\partial u^3}\vec{i} + \frac{\partial y}{\partial u^3}\vec{j} + \frac{\partial z}{\partial u^3}\vec{k}\right)\right]$$

$$= \frac{\partial \vec{r}}{\partial u^1} \cdot \left(\frac{\partial \vec{r}}{\partial u^2} \times \frac{\partial \vec{r}}{\partial u^3}\right) = h_1\vec{e}_1 \cdot (h_2\vec{e}_2 \times h_3\vec{e}_3) = h_1h_2h_3\left[\vec{e}_1 \cdot (\vec{e}_2 \times \vec{e}_3)\right]$$

$$= h_1h_2h_3\,(\vec{e}_1 \cdot \vec{e}_1) = h_1h_2h_3$$

If $J \equiv 0$ then there is a relation between x, y, z having the form $F(x, y, z) = 0$, that is the coordinate transformation breaks down. We shall therefore require that $J \neq 0$. If (u^1, u^2, u^3) are general curvilinear coordinates in Euclidean 3-space then, from calculus

$$\frac{\partial \vec{r}}{\partial u^1}, \frac{\partial \vec{r}}{\partial u^2}, \frac{\partial \vec{r}}{\partial u^3} \quad \text{and} \quad \nabla u^1, \nabla u^2, \nabla u^3$$

are *reciprocal* systems of vectors, that is

$$\frac{\partial \vec{r}}{\partial u^p} \cdot \nabla u^q = \begin{cases} 1 & \text{if} \quad p = q \\ 0 & \text{if} \quad p \neq q \end{cases}$$

It's not to be wondered at that $\dfrac{\partial\vec{r}}{\partial u^p}$ and ∇u^p are called *unitary* base vectors.

Also, from calculus, if $\vec{A}, \vec{B}, \vec{C}$ and $\bar{\vec{A}}, \bar{\vec{B}}, \bar{\vec{C}}$ are reciprocal sets of vectors and $\vec{A}\cdot(\vec{B}\times\vec{C}) = V$ then $\bar{\vec{A}}\cdot(\bar{\vec{B}}\times\bar{\vec{C}}) = 1/V$. From the above results we have

$$\left[\frac{\partial\vec{r}}{\partial u^1}\cdot\left(\frac{\partial\vec{r}}{\partial u^2}\times\frac{\partial\vec{r}}{\partial u^3}\right)\right]\left[\nabla u^1\cdot(\nabla u^2\times\nabla u^3)\right]=1$$

We have just seen that if (u^1, u^2, u^3) is general curvilinear coordinates, then

$$J\left(\frac{x, y, z}{u^1, u^2, u^3}\right)=\frac{\partial\vec{r}}{\partial u^1}\cdot\left(\frac{\partial\vec{r}}{\partial u^2}\times\frac{\partial\vec{r}}{\partial u^3}\right)$$

But

$$J\left(\frac{u^1, u^2, u^3}{x, y, z}\right)=\begin{vmatrix}\dfrac{\partial u^1}{\partial x} & \dfrac{\partial u^1}{\partial y} & \dfrac{\partial u^1}{\partial z}\\ \dfrac{\partial u^2}{\partial x} & \dfrac{\partial u^2}{\partial y} & \dfrac{\partial u^2}{\partial z}\\ \dfrac{\partial u^3}{\partial x} & \dfrac{\partial u^3}{\partial y} & \dfrac{\partial u^3}{\partial z}\end{vmatrix}=\nabla u^1\cdot(\nabla u^2\times\nabla u^3)$$

Therefore

$$J\left(\frac{x, y, z}{u^1, u^2, u^3}\right)J\left(\frac{u^1, u^2, u^3}{x, y, z}\right)=1$$

33. Area and volume in Euclidean 3-space

If (u^1, u^2, u^3) are general curvilinear coordinates in Euclidean 3-space and $\vec{r} = \vec{r}(u^1, u^2, u^3)$ is the position vector of a point *P*, we have

$$d\vec{r} = \frac{\partial \vec{r}}{\partial u^1} du^1 + \frac{\partial \vec{r}}{\partial u^2} du^2 + \frac{\partial \vec{r}}{\partial u^3} du^3$$

$$= h_1 du^1 \vec{e}_1 + h_2 du^2 \vec{e}_2 + h_3 du^3 \vec{e}_3$$

For orthogonal systems, $\vec{e}_1 \cdot \vec{e}_2 = \vec{e}_2 \cdot \vec{e}_3 = \vec{e}_3 \cdot \vec{e}_1 = 0$ so that

$$ds^2 = d\vec{r} \cdot d\vec{r} = h_1^2 \left(du^1\right)^2 + h_2^2 \left(du^2\right)^2 + h_3^2 \left(du^3\right)^2$$

Note that in this case

$$g_{11} = h_1^2, \quad g_{22} = h_2^2, \quad g_{33} = h_3^2$$

Along a u^1 curve, u^2 and u^3 are constants and so $d\vec{r} = h_1 du^1 \vec{e}_1$. The differential of arc length ds_1 along the u^1 curve at *P* is

$$ds_1 = \left|d\vec{r}\right| = \left|h_1 du^1 \vec{e}_1\right| = h_1 du^1$$

Similarly, $ds_2 = h_2 du^2, \quad ds_3 = h_3 du^3$ at *P*. The *elements of area* in orthogonal curvilinear coordinates are therefore

$$dA_1 = ds_2 ds_3 = h_2 h_3 du^2 du^3$$
$$dA_2 = ds_1 ds_3 = h_1 h_3 du^1 du^3$$
$$dA_3 = ds_1 ds_2 = h_1 h_2 du^1 du^2$$

Another method (from calculus):

$$dA_1 = \left|\left(h_2 du^2 \vec{e}_2\right) \times \left(h_3 du^3 \vec{e}_3\right)\right| = h_2 h_3 du^2 du^3$$

$$dA_2 = \left|\left(h_1 du^1 \vec{e}_1\right) \times \left(h_3 du^3 \vec{e}_3\right)\right| = h_1 h_3 du^1 du^3$$

$$dA_3 = \left|\left(h_1 du^1 \vec{e}_1\right) \times \left(h_2 du^2 \vec{e}_2\right)\right| = h_1 h_2 du^1 du^2$$

Obviously, for rectangular Cartesian coordinates (*x*, *y*, *z*) we have $h_1 = h_2 = h_3 = 1$ so that

$$\begin{aligned} dA_1 &= dydz \\ dA_2 &= dxdz \\ dA_3 &= dxdy \end{aligned}$$

The *volume element* for a right-handed orthogonal curvilinear coordinate system is

$$dV = ds_1 ds_2 ds_3 = h_1 h_2 h_3 du^1 du^2 du^3 = J\left(\frac{x, y, z}{u^1, u^2, u^3}\right) du^1 du^2 du^3$$

Another method (from calculus):

$$\begin{aligned} dV &= \left|\left(h_1 du^1 \vec{e}_1\right) \cdot \left[\left(h_2 du^2 \vec{e}_2\right) \times \left(h_3 du^3 \vec{e}_3\right)\right]\right| = h_1 h_2 h_3 du^1 du^2 du^3 \\ &= J\left(\begin{matrix} x, y, z \\ u^1, u^2, u^3 \end{matrix}\right) du^1 du^2 du^3 \end{aligned}$$

Obviously, for rectangular Cartesian coordinates (*x*, *y*, *z*) we have $h_1 = h_2 = h_3 = 1$ so that $dV = dxdydz$.

In general curvilinear coordinates (u^1, u^2, u^3) in Euclidean 3-space the volume element is given by

$$dV = \sqrt{g}\, du^1 du^2 du^3$$

where

$$g = \begin{vmatrix} g_{11} & g_{12} & g_{13} \\ g_{21} & g_{22} & g_{23} \\ g_{31} & g_{32} & g_{33} \end{vmatrix}$$

Proof

If $\vec{r} = \vec{r}(u^1, u^2, u^3)$ is the position vector of a point *P*, we have

$$d\vec{r} = \frac{\partial \vec{r}}{\partial u^1} du^1 + \frac{\partial \vec{r}}{\partial u^2} du^2 + \frac{\partial \vec{r}}{\partial u^3} du^3$$

Along a u^1 curve, u^2 and u^3 are constants and so $d\vec{r} = \frac{\partial \vec{r}}{\partial u^1} du^1$. Similarly $d\vec{r} = \frac{\partial \vec{r}}{\partial u^2} du^2$ and $d\vec{r} = \frac{\partial \vec{r}}{\partial u^3} du^3$ at *P*. The volume element is therefore

$$dV = \left| \left(\frac{\partial \vec{r}}{\partial u^1} du^1 \right) \cdot \left[\left(\frac{\partial \vec{r}}{\partial u^2} du^2 \right) \times \left(\frac{\partial \vec{r}}{\partial u^3} du^3 \right) \right] \right| = \left| \frac{\partial \vec{r}}{\partial u^1} \cdot \left(\frac{\partial \vec{r}}{\partial u^2} \times \frac{\partial \vec{r}}{\partial u^3} \right) \right| du^1 du^2 du^3$$

Now we must show that

$$\sqrt{g} = \left| \frac{\partial \vec{r}}{\partial u^1} \cdot \left(\frac{\partial \vec{r}}{\partial u^2} \times \frac{\partial \vec{r}}{\partial u^3} \right) \right|$$

or

$$g = \left[\frac{\partial \vec{r}}{\partial u^1} \cdot \left(\frac{\partial \vec{r}}{\partial u^2} \times \frac{\partial \vec{r}}{\partial u^3} \right) \right]^2$$

We have (from calculus)

$$g_{pq} = \frac{\partial \vec{r}}{\partial u^p} \cdot \frac{\partial \vec{r}}{\partial u^q} = \frac{\partial x}{\partial u^p}\frac{\partial x}{\partial u^q} + \frac{\partial y}{\partial u^p}\frac{\partial y}{\partial u^q} + \frac{\partial z}{\partial u^p}\frac{\partial z}{\partial u^q}$$

Then, using the following theorem on multiplication of determinants,

$$\begin{vmatrix} a_1 & a_2 & a_3 \\ b_1 & b_2 & b_3 \\ c_1 & c_2 & c_3 \end{vmatrix}\begin{vmatrix} A_1 & B_1 & C_1 \\ A_2 & B_2 & C_2 \\ A_3 & B_3 & C_3 \end{vmatrix}$$

$$= \begin{vmatrix} a_1A_1 + a_2A_2 + a_3A_3 & a_1B_1 + a_2B_2 + a_3B_3 & a_1C_1 + a_2C_2 + a_3C_3 \\ b_1A_1 + b_2A_2 + b_3A_3 & b_1B_1 + b_2B_2 + b_3B_3 & b_1C_1 + b_2C_2 + b_3C_3 \\ c_1A_1 + c_2A_2 + c_3A_3 & c_1B_1 + c_2B_2 + c_3B_3 & c_1C_1 + c_2C_2 + c_3C_3 \end{vmatrix}$$

we have

$$\left[\frac{\partial \vec{r}}{\partial u^1} \cdot \left(\frac{\partial \vec{r}}{\partial u^2} \times \frac{\partial \vec{r}}{\partial u^3}\right)\right]^2 = \begin{vmatrix} \frac{\partial x}{\partial u^1} & \frac{\partial y}{\partial u^1} & \frac{\partial z}{\partial u^1} \\ \frac{\partial x}{\partial u^2} & \frac{\partial y}{\partial u^2} & \frac{\partial z}{\partial u^2} \\ \frac{\partial x}{\partial u^3} & \frac{\partial y}{\partial u^3} & \frac{\partial z}{\partial u^3} \end{vmatrix}^2$$

$$= \begin{vmatrix} \frac{\partial x}{\partial u^1} & \frac{\partial y}{\partial u^1} & \frac{\partial z}{\partial u^1} \\ \frac{\partial x}{\partial u^2} & \frac{\partial y}{\partial u^2} & \frac{\partial z}{\partial u^2} \\ \frac{\partial x}{\partial u^3} & \frac{\partial y}{\partial u^3} & \frac{\partial z}{\partial u^3} \end{vmatrix}\begin{vmatrix} \frac{\partial x}{\partial u^1} & \frac{\partial y}{\partial u^1} & \frac{\partial z}{\partial u^1} \\ \frac{\partial x}{\partial u^2} & \frac{\partial y}{\partial u^2} & \frac{\partial z}{\partial u^2} \\ \frac{\partial x}{\partial u^3} & \frac{\partial y}{\partial u^3} & \frac{\partial z}{\partial u^3} \end{vmatrix}$$

$$= \begin{vmatrix} g_{11} & g_{12} & g_{13} \\ g_{21} & g_{22} & g_{23} \\ g_{31} & g_{32} & g_{33} \end{vmatrix} = g$$

or

$$\sqrt{g} = \left| \frac{\partial \vec{r}}{\partial u^1} \cdot \left(\frac{\partial \vec{r}}{\partial u^2} \times \frac{\partial \vec{r}}{\partial u^2} \right) \right|$$

Note that

$$\sqrt{g} = \left| J\left(\frac{x, y, z}{u^1, u^2, u^3} \right) \right|$$

where (u^1, u^2, u^3) are general curvilinear coordinates. But for right-handed orthogonal coordinate systems

$$\sqrt{g} = J\left(\frac{x, y, z}{u^1, u^2, u^3} \right)$$

dV is an invariant and $\sqrt{g}$ is a *relative tensor of weight one* (see other books). Let V denote a closed region in an Euclidean 3-space. Then the volume of the region is

$$\iiint_V dV$$

When transforming multiple integrals from right-handed rectangular coordinates to right-handed orthogonal curvilinear coordinates the volume element *dxdydz* is replaced by

$$h_1 h_2 h_3 du^1 du^2 du^3 \quad \text{or} \quad J\left(\frac{x, y, z}{u^1, u^2, u^3} \right) du^1 du^2 du^3$$

When transforming to general coordinates the volume element *dxdydz* is replaced by

$$\sqrt{g}du^1du^2du^3 \quad \text{or} \quad \left|J\left(\frac{x,y,z}{u^1,u^2,u^3}\right)\right|du^1du^2du^3$$

34. The Kronecker delta (the Kronecker tensor)

The *Kronecker delta* or the *Kronecker tensor*, written δ_q^p, and defined by

$$\delta_q^p = \begin{cases} 0 & \text{if } p \neq q \\ 1 & \text{if } p = q \end{cases}$$

is a mixed tensor of rank 2, justifying the notation used.

Proof

We must prove that

$$\bar{\delta}_k^j = \frac{\partial \bar{u}^j}{\partial u^p} \frac{\partial u^q}{\partial \bar{u}^k} \delta_q^p$$

By definition

$$\bar{\delta}_q^p = \delta_q^p = \begin{cases} 0 & \text{if } p \neq q \\ 1 & \text{if } p = q \end{cases}$$

Coordinates $\bar{u}^j$ are functions of coordinates u^p which are in turn functions of coordinates $\bar{u}^k$. Then, by the chain rule,

$$\frac{\partial \bar{u}^j}{\partial u^p} \frac{\partial u^p}{\partial \bar{u}^k} = \frac{\partial \bar{u}^j}{\partial \bar{u}^k} = \delta_k^j$$

Now we must prove that the right side of the transformation equations of the mixed tensor above equals $\bar{\delta}_k^j$. It does:

$$\frac{\partial \bar{u}^j}{\partial u^p} \frac{\partial u^q}{\partial \bar{u}^k} \delta_q^p = \frac{\partial \bar{u}^j}{\partial u^p} \frac{\partial u^p}{\partial \bar{u}^k} = \frac{\partial \bar{u}^j}{\partial \bar{u}^k} = \delta_k^j = \bar{\delta}_k^j$$

The Kronecker delta is a *constant* or *numerical tensor*, that is, it has the same components in all coordinate systems. Aside from the scalars and zero, δ^p_q (together with its direct products) is the only tensor whose components are the same in all coordinate systems.

35. The conjugate metric tensor (the reciprocal metric tensor)

The metric determinant of a Riemannian N-space is

$$g = \begin{vmatrix} g_{11} & g_{12} & \cdot & \cdot & \cdot & g_{1N} \\ g_{21} & g_{22} & \cdot & \cdot & \cdot & g_{2N} \\ \cdot & \cdot & & & & \cdot \\ \cdot & \cdot & & & & \cdot \\ \cdot & \cdot & & & & \cdot \\ g_{N1} & g_{N2} & \cdot & \cdot & \cdot & g_{NN} \end{vmatrix}$$

with elements g_{pq}. Now we define

$$g^{jk} = \frac{\text{cofactor of } g_{jk}}{g}$$

We suppose that $g \neq 0$. g is, of course, a scalar function, depending on the coordinate system. Therefore, g is not an invariant. In a Riemannian 3-space, if

$$g = \begin{vmatrix} g_{11} & g_{12} & g_{13} \\ g_{21} & g_{22} & g_{23} \\ g_{31} & g_{32} & g_{33} \end{vmatrix}$$

we have from calculus

$$G(2,1) = \text{cofactor of } g_{21} = (-1)^{2+1} \begin{vmatrix} g_{12} & g_{13} \\ g_{32} & g_{33} \end{vmatrix}$$

$$G(2,2) = \text{cofactor of } g_{22} = (-1)^{2+2} \begin{vmatrix} g_{12} & g_{13} \\ g_{31} & g_{33} \end{vmatrix}$$

$$G(2,3) = \text{cofactor of } g_{23} = (-1)^{2+3} \begin{vmatrix} g_{11} & g_{12} \\ g_{31} & g_{32} \end{vmatrix}$$

Then, by a theorem of determinants,

$$g_{21}G(2,1)+g_{22}G(2,2)+g_{23}G(2,3)=g$$

More generally, it can be shown that if g is an Nth order determinant, then

$$g_{jk}G(j,k)=g$$

where summation is over k only.
By a theorem of determinants

$$g_{21}G(3,1)+g_{22}G(3,2)+g_{23}G(3,3)=0$$

More generally, it can be shown that if g is an Nth order determinant, then

$$g_{jk}G(p,k)=0 \quad \text{if} \quad p\neq j$$

where summation is over k only. Then we have the following results

$$g_{jk}\frac{G(j,k)}{g}=1$$

where summation is over k only, and

$$g_{jk}\frac{G(p,k)}{g}=0 \quad \text{if} \quad p\neq j$$

where summation is over k only. By definition (see above)

$$g^{jk}=\frac{G(j,k)}{g}$$

so we have

$$g_{jk}g^{jk} = 1$$

where summation is over k only (if not, then $g_{jk}g^{jk} = N$), and

$$g_{jk}g^{pk} = 0 \quad \text{if} \quad p \neq j$$

where summation is over k only. Then

$$g_{jk}g^{pk} = \delta_j^p$$

$G(j,k)$ is symmetric because g_{jk} is symmetric, and g^{jk} is symmetric because $G(j,k)$ is symmetric. Thus, by the quotient law, g^{pq} is a symmetric contravariant tensor of rank 2. It is called the *conjugate metric tensor.* g_{pq} and g^{pq} are said to be *conjugate* to one another. Note that $g_{pq} = g^{pq} = \delta_q^p$ in Euclidean *N*-spaces when the metric form is transformed into $ds^2 = dx^k dx^k$.

36. Associated tensors. Raising and lowering indices

In Euclidean N-space with orthogonal coordinates u^r and $\overline{u}^r$, we have

$$\overline{C}^p = \frac{\partial \overline{u}^p}{\partial u^q} C^q$$

and

$$\overline{C}_p = \frac{\partial u^q}{\partial \overline{u}^p} C_q$$

where C^p and C_p

are the contravariant and covariant components of the same geometrical object. However, in general N-space the terms C^p and C_p are entirely unrelated; one is contravariant and the other covariant, and there is no connection between the one and the other. The use of the same letter C in both implies no relationship. But it is shown that in a general Riemannian N-space with general coordinates, these two tensors are the same geometrical object. And it is shown that if we know the components of one, we can find the components of the other. In a general Riemannian N-space with general coordinates, a metric tensor g_{pq} (and a conjugate metric tensor g^{pq}), we have, for example

$$C^q = g^{pq} C_p$$

$$C_q = g_{pq} C^p$$

For tensors in an Euclidean N-space with orthogonal coordinates x^r and $\overline{x}^r$, and where x^r is Cartesian coordinates, we have (for example):

$$\text{Given} \quad \bar{C}_q = \frac{\partial x^p}{\partial \bar{x}^q} C_p$$

$$\text{then} \quad \bar{C}_q = \frac{\partial x^p}{\partial \bar{x}^q} C^p$$

$$\text{or} \quad \bar{C}_q = \frac{\partial x^1}{\partial \bar{x}^q} C^1 + \frac{\partial x^2}{\partial \bar{x}^q} C^2 + \ldots + \frac{\partial x^N}{\partial \bar{x}^q} C^N$$

$$\text{or} \quad \bar{C}_q = \frac{\partial x^1}{\partial \bar{x}^q} \frac{\partial x^1}{\partial \bar{x}^p} \bar{C}^p + \frac{\partial x^2}{\partial \bar{x}^q} \frac{\partial x^2}{\partial \bar{x}^p} \bar{C}^p + \ldots + \frac{\partial x^N}{\partial \bar{x}^q} \frac{\partial x^N}{\partial \bar{x}^p} \bar{C}^p$$

$$\text{or} \quad \bar{C}_q = \frac{\partial x^1}{\partial \bar{x}^p} \frac{\partial x^1}{\partial \bar{x}^q} \bar{C}^p + \frac{\partial x^2}{\partial \bar{x}^p} \frac{\partial x^2}{\partial \bar{x}^q} \bar{C}^p + \ldots + \frac{\partial x^N}{\partial \bar{x}^p} \frac{\partial x^N}{\partial \bar{x}^q} \bar{C}^p$$

$$\text{or} \quad \bar{C}_q = \left(\frac{\partial x^1}{\partial \bar{x}^p} \frac{\partial x^1}{\partial \bar{x}^q} + \frac{\partial x^2}{\partial \bar{x}^p} \frac{\partial x^2}{\partial \bar{x}^q} + \ldots + \frac{\partial x^N}{\partial \bar{x}^p} \frac{\partial x^N}{\partial \bar{x}^q} \right) \bar{C}^p$$

$$\text{or} \quad \bar{C}_q = \bar{g}_{pq} \bar{C}^p$$

$$\text{or} \quad C_q = g_{pq} C^p$$

A general Riemannian *N*-space with general coordinates x^r can always be considered as immersed in a Euclidean N-space with Cartesian coordinates $y^k = y^k(x^1, x^2, \ldots, x^N)$. (We omit the proof.) We determine the *N* functions $y^k = y^k(x^1, x^2, \ldots, x^N)$ such that when we differentiate them and take the sum of the squares of the differentials we get a form, quadratic in the dx^r, which is identical with the given ds^2, so that we have identically

$$ds^2 = dy^k dy^k = g_{pq} dx^p dx^q$$

Expressing the dy^k in terms of the dx^r, we have

$$\frac{\partial y^k}{\partial x^p} \frac{\partial y^k}{\partial x^q} dx^p dx^q = g_{pq} dx^p dx^q$$

Or, equating the coefficients of $dx^p dx^q$, we have

$$g_{pq} = \frac{\partial y^k}{\partial x^p} \frac{\partial y^k}{\partial x^q}$$

Using the result above, it can be proved that the process of raising and lowering indices holds quite generally: We can lower or raise indices by forming inner products of the given tensor with the metric tensor g_{pq} or the conjugate metric tensor g^{pq}, respectively.

All tensors obtained from a given tensor by forming inner products with the metric tensor and/or its conjugate are called *associated tensors* of the given tensor. For example, A^p and A_p are associated tensors, the first are the contravariant components and the second are the covariant components of the same geometrical object: the tensor with A as the *principal letter*.

We shall often in the future refrain from writing a subscript and a superscript on the same vertical line. In vacant spaces we shall write dots. Example:

$$A_{nrs} = g_{nm} A^{m}_{\bullet rs}$$

In the case of non-symmetrical tensors it may be necessary to distinguish the place from which the raised suffix has been brought, e.g. to distinguish between $A_m^{\bullet n}$ and $A^{n}_{\bullet m}$. Where no confusion can arise we can omit the dots.

Examples:

$$A^{p}_{\bullet q} = g^{rp} A_{rq}$$

$$A_r^{\bullet q} = g^{sq} A_{rs}$$

$$A^{pq} = g^{rp} A_r^{\bullet q}$$

$$A^{pq} = g^{rp} g^{sq} A_{rs}$$

$$A^{p}_{\bullet rs} = g_{rq} A^{pq}_{\bullet\bullet s}$$

$$A^{qmst}_{\bullet\bullet\bullet\bullet p} = g^{rm} A^{q\bullet st}_{\bullet r\bullet\bullet p}$$

$$A^{qm\bullet t}_{\bullet\bullet n\bullet p} = g_{sn} A^{qmst}_{\bullet\bullet\bullet\bullet p}$$

$$A^{qm\bullet tk}_{\bullet\bullet n\bullet} = g^{pk} A^{qm\bullet t}_{\bullet\bullet n\bullet p}$$

$$A^{qm\bullet tk}_{\bullet\bullet n} = g^{pk} g_{sn} g^{rm} A^{q\bullet st}_{\bullet r\bullet\bullet p}$$

Note that inner multiplication by δ^{m}_{n} gives *plain substitution*. Examples:

$$A^{m} = \delta^{m}_{n} A^{n}$$

$$T^{m}_{\bullet rs} = g^{mn} g_{nm} T^{m}_{\bullet rs}$$

37. The physical components of a vector A^p or A_p

What are usually called the components of a vector $\vec{A}$ in elementary treatments are not the contravararant components A^p or the covariant components A_p, but the components $\bar{A}_1, \bar{A}_2, \bar{A}_3$ of $\vec{A} = \bar{A}_1\vec{e}_1 + \bar{A}_2\vec{e}_2 + \bar{A}_3\vec{e}_3$.

In orthogonal curvilinear coordinates in Euclidean 3-space we have

$$\begin{aligned}\bar{A}_1 &= \sqrt{g_{11}}\,A^1\\ \bar{A}_2 &= \sqrt{g_{22}}\,A^2\\ \bar{A}_3 &= \sqrt{g_{33}}\,A^3\end{aligned}$$

Proof

In this proof p,q = 1,2,3, of course. We know that

$$\frac{\partial \vec{r}}{\partial u^p} = h_p\vec{e}_p, \quad \frac{\partial \vec{r}}{\partial u^q} = h_q\vec{e}_q \quad \text{(no summation)}$$

Then

$$g_{pq} = \frac{\partial \vec{r}}{\partial u^p} \cdot \frac{\partial \vec{r}}{\partial u^q} = h_p\vec{e}_p \cdot h_q\vec{e}_q = h_p h_q\left(\vec{e}_p \cdot \vec{e}_q\right) = h_p h_q \quad \text{(no summation)}$$

so

$$g_{pp} = h_p^{\ 2} \quad \text{or} \quad h_p = \sqrt{g_{pp}} \quad \text{(no summation)}$$

Then we have

$$\vec{A} = A^p \frac{\partial \vec{r}}{\partial u^p} = A^p\sqrt{g_{pp}}\,\vec{e}_p = \sqrt{g_{pp}}\,A^p\vec{e}_p \quad \text{(summation over } A^p\vec{e}_p \text{ only)}$$

and now this proof is completed.

Also, in orthogonal curvilinear coordinates in Euclidean 3-space we have

$$\bar{A}_1 = \frac{A_1}{\sqrt{g_{11}}}$$

$$\bar{A}_2 = \frac{A_2}{\sqrt{g_{22}}}$$

$$\bar{A}_3 = \frac{A_1}{\sqrt{g_{33}}}$$

Proof

In this proof p,q = 1,2,3, of course. From section 36 we have $A^p = g^{pq}A_q$ so $A^1 = g^{11}A_1 + g^{12}A_2 + g^{13}A_3 = g^{11}A_1$. Similarly, $A^2 = g^{22}A_2$, $A^3 = g^{33}A_3$. In orthogonal curvilinear coordinates in "ordinary" Euclidean 3-space we have

$$g^{pp} = \frac{1}{g_{pp}} \quad \text{or} \quad \sqrt{g^{pp}} = \frac{1}{\sqrt{g_{pp}}} \quad \text{(no summation)}$$

Therefore

$$\bar{A}_1 = \sqrt{g_{11}}\,A^1 = \frac{1}{\sqrt{g^{11}}}g^{11}A_1 = \frac{\sqrt{g^{11}}\sqrt{g^{11}}}{\sqrt{g^{11}}}A_1 = \sqrt{g^{11}}A_1 = \frac{A_1}{\sqrt{g_{11}}}$$

Similarly, $\bar{A}_2 = \dfrac{A_2}{\sqrt{g_{22}}}$, $\bar{A}_3 = \dfrac{A_3}{\sqrt{g_{33}}}$.

38. The physical components of a tensor A^{pq} or A_{pq}

Not every tensor can be written as a product of two tensors of lower rank. But let us here suppose that, in orthogonal curvilinear coordinates in Euclidean 3-space,

$$A^{pq} = B^p B^q, \quad A_{pq} = B_p B_q$$

Then the physical components are expressed in terms of contravariant components as follows

$$\begin{array}{lll}
\overline{A}_{11} = \sqrt{g_{11}g_{11}}\,A^{11} & \overline{A}_{11} = \sqrt{g_{11}g_{22}}\,A^{12} & \overline{A}_{13} = \sqrt{g_{11}g_{33}}\,A^{13} \\
\overline{A}_{21} = \sqrt{g_{22}g_{11}}\,A^{21} & \overline{A}_{22} = \sqrt{g_{22}g_{22}}\,A^{22} & \overline{A}_{23} = \sqrt{g_{22}g_{33}}\,A^{23} \\
\overline{A}_{31} = \sqrt{g_{33}g_{11}}\,A^{31} & \overline{A}_{32} = \sqrt{g_{33}g_{22}}\,A^{32} & \overline{A}_{33} = \sqrt{g_{33}g_{33}}\,A^{33}
\end{array}$$

And the physical components are expressed in terms of covariant components as follows

$$\begin{array}{lll}
\overline{A}_{11} = \dfrac{A_{11}}{\sqrt{g_{11}g_{11}}} & \overline{A}_{12} = \dfrac{A_{12}}{\sqrt{g_{11}g_{22}}} & \overline{A}_{13} = \dfrac{A_{13}}{\sqrt{g_{11}g_{33}}} \\
\overline{A}_{21} = \dfrac{A_{21}}{\sqrt{g_{22}g_{11}}} & \overline{A}_{22} = \dfrac{A_{22}}{\sqrt{g_{22}g_{22}}} & \overline{A}_{23} = \dfrac{A_{23}}{\sqrt{g_{22}g_{33}}} \\
\overline{A}_{31} = \dfrac{A_{31}}{\sqrt{g_{33}g_{11}}} & \overline{A}_{32} = \dfrac{A_{32}}{\sqrt{g_{33}g_{22}}} & \overline{A}_{33} = \dfrac{A_{33}}{\sqrt{g_{33}g_{33}}}
\end{array}$$

Proof

In this proof p,q = 1,2,3, of course.

$$\overline{A}_{pq} = \overline{B}_p \overline{B}_q = \left(\sqrt{g_{pp}}\,B^p\right)\left(\sqrt{g_{qq}}\,B^q\right) = \sqrt{g_{pp}g_{qq}}\,A^{pq} \quad \text{(no summation)}$$

and

$$\bar{A}_{pq} = \bar{B}_p \bar{B}_q = \frac{B_p}{\sqrt{g_{pp}}} \frac{B_q}{\sqrt{g_{qq}}} = \frac{A_{pq}}{\sqrt{g_{pp} g_{qq}}} \qquad \text{(no summation)}$$

and now this proof is completed.

39. Distance in a Riemannian space

The *distance s* between two points $u^r(t_1)$ and $u^r(t_2)$ on a curve $u^r = u^r(t)$, where t is a parameter, in a general Riemannian *N*-space with general coordinates u^r, is given by

$$s = \int_{t_1}^{t_2} \sqrt{g_{pq} \frac{du^p}{dt} \frac{du^q}{dt}} dt$$

Proof

Given $ds^2 = g_{pq} du^p du^q$, then $ds = \sqrt{g_{pq} du^p du^q}$ and $\frac{ds}{dt} = \sqrt{g_{pq} \frac{du^p}{dt} \frac{du^q}{dt}}$. From calculus $s = \int_{t_1}^{t_2} ds$ or $s = \int_{t_1}^{t_2} \frac{ds}{dt} dt$. Then we have

$$s = \int_{t_1}^{t_2} \sqrt{g_{pq} \frac{du^p}{dt} \frac{du^q}{dt}} dt.$$

That curve in the space which makes the distance s a minimum is called a geodesic (minimizing geodesic) of the space.

Appendix:

In Euclidean spaces the metric form $g_{pq} du^p du^q$ is *positive-definite.* This means that it is positive unless all the differentials (du^r) vanish. In other words, the distance between two points vanishes only if the points are brought into coincidence. In general Riemannian spaces, of course, if the metric form is negative, the *ds* must be imaginary. The mathematics of relativity must admit the possibility of an *indefinite* metric form such as

$$\left(du^1\right)^2 - \left(du^2\right)^2$$

For some displacements du^r an indefinite metric form may be positive and for others it may be zero or negative. Note that in a Riemannian space with an indefinite metric form, two points may be at zero distance from one another without being coincident. It is important to note that Riemannian geometry is built up on the concept of the distance between two adjacent points, or more precisely the differentials du^r, rather than on the concept of finite distance.
To overcome the difficulty in the definition of distance arising from the indefiniteness of the metric form, we can define the *length* of the displacement du^r to be *ds*, where

$$ds^2 = \varepsilon\Phi = \varepsilon g_{pq} du^p du^q, \quad ds \geq 0$$

Here $\Phi = g_{pq} du^p du^q$ is the metric form and ε is an *indicator* chosen equal to +1, or −1, so as to make $\varepsilon\Phi$ positive if Φ is negative (or positive). If $\Phi = 0$, for du^r not all zero, the displacement is called a *null displacement*. As we see, we define the length of a null displacement to be zero.
Obviously, we can now define

$$s = \int_{t_1}^{t_2} \sqrt{\varepsilon g_{pq} \frac{du^p}{dt} \frac{du^q}{dt}} dt$$

where ε is the indicator (see above) so as to make *ds* positive. If $s = 0$ ($ds = 0$) along the curve, then the curve is called a *null curve*.

40. Angle between two curves in a Riemannian space

Consider Euclidean 3-space. The *angle* between two curves in the space is the same thing as the angle between two vectors, tangent to the curves. We want to set up a *definition* of angle in "ordinary" Euclidean 3-space in such a way that it can be taken over into Riemannian spaces. The *law of cosines* for plane triangles can be written

$$BC^2 = AB^2 + AC^2 - 2\ AB\ AC \cos\theta \qquad \text{or} \qquad \cos\theta = \frac{AB^2 + AC^2 - BC^2}{2\ AB\ AC}$$

It is easy to see that the angle θ between two curves through the points *A* and *B*, and *A* and *C*, respectively, in Euclidean 3-space satisfies the equation

$$\cos\theta = \lim_{\substack{AB \to 0 \\ AC \to 0}} \frac{AB^2 + AC^2 - BC^2}{2\ AB\ AC}$$

where, of course, *AB*, *AC*, and *BC* are arc-lengths.

Now suppose that *A*, *B*, and *C* are points in a Riemannian N-space with a positive-definite metric. Then, taking the limit as *AB* and *AC* approach zero, it is easy to see that

$$AB^2 \to d_B s^2 = g_{pq} d_B x^p d_B x^q$$
$$AC^2 \to d_C s^2 = g_{pq} d_C x^p d_C x^q$$
$$BC^2 \to d_{BC} s^2 = g_{pq} d_{BC} x^p d_{BC} x^q = g_{pq} \left(d_C x^p - d_B x^p\right)\left(d_C x^q - d_B x^q\right)$$

where g_{pq} is evaluated at A because A is at the point of the limit itself. Note the use of the *law for addition/subtraction of vectors*. Now we see that

$$\cos\theta = \lim \frac{AB^2 + AC^2 - BC^2}{2\,AB\,AC}$$

$$AB \to 0$$

$$AC \to 0$$

$$= \frac{g_{pq} d_B x^p d_B x^q + g_{pq} d_C x^p d_C x^q - g_{pq}\left(d_C x^p - d_B x^q\right)\left(d_C x^p - d_B x^q\right)}{2\, d_B s\, d_C s}$$

$$= \frac{g_{pq} d_B x^p d_C x^q}{d_B s\, d_C s} = g_{pq} \frac{d_B x^p}{d_B s} \frac{d_C x^q}{d_C s}$$

where $\frac{d_B x^p}{d_B s}$ and $\frac{d_C x^q}{d_C s}$ are unit tangent vectors to the two curves at A, respectively. Hence the angle θ between two curves (or between any two unit vectors at a point) in a general Riemannian N-space with a positive-definite metric and general coordinates satisfies

$$\cos\theta = g_{pq} \frac{d_B x^p}{d_B s} \frac{d_C x^q}{d_C s}$$

It can be shown that the right hand side of the above equation does not exceed unity in absolute value in a Riemannian space with a positive-definite metric. It can also be shown that this equation determines a unique θ in the range (0, π).

41. Length of a vector in a Riemannian space

Recall from calculus that

$$\vec{A} \square \vec{B} = A_1 B_1 + A_2 B_2 + A_3 B_3 = A_p B_p \quad (p = 1, 2, 3, \text{ summed })$$

$$\vec{A} \square \vec{A} = A_1 A_1 + A_2 A_2 + A_3 A_3 = L^2 = A_p A_p \quad (p = 1, 2, 3, \text{ summed})$$

where L is the length (magnitude) of $\vec{A}$. Then, clearly, the quantity $A^p B_p$, which is the inner product of A^p and B_q, is a scalar analogous to the scalar product in rectangular Cartesian coordinates in "ordinary" Euclidean 3-space. Therefore we define the length L of the vector A^p (or A_p) in a general Riemannian N-space with general coordinates as given by

$$L^2 = A^p A_p = A_p A^p = g_{pq} A^p A^q = g^{pq} A_p A_q$$

or

$$L = \sqrt{A^p A_p} = \sqrt{A_p A^p} = \sqrt{g_{pq} A^p A^q} = \sqrt{g^{pq} A_p A_q}$$

Clearly L is an invariant.

42. Angle between two vectors in a Riemannian space

Recall from calculus, and section 41, that

$$\vec{A}\cdot\vec{B} = AB\cos\theta \quad \text{or} \quad \cos\theta = \frac{\vec{A}\cdot\vec{B}}{AB} = \frac{A_p B_p}{AB}, \quad (p = 1, 2, 3, \text{summed})$$

Therefore we define the angle θ between A^p and B_p in a general Riemannian N-space with general coordinates as given by

$$\cos\theta = \frac{A^p B_p}{\sqrt{\left(A^p A_p\right)\left(B^p B_p\right)}}$$

Clearly θ and $\cos\theta$ are invariants.

43. Notes on spaces, coordinates, tensors, and physics

Most physicists are satisfied with an intuitive definition of a differentiable mathematical space (differentiable manifold): A differentiable space is a set of "points" tied together continuously and differentiably, so that the points in any *sufficiently small* region of physical space can be put into a one-to-one correspondence with a set of points of the *number space* of ordered N-tuples $(u^1, u^2, \ldots, u^N)$. That correspondence furnishes a coordinate system for the neighborhood. The differentiability of a space is, of course, the possibility of defining differentiable functions on it. Differentiability permits one to *introduce* coordinate systems locally, if not globally, and also permits one to *introduce* curves, tangent vectors (tensors of rank one), tensors, etc. But the mere fact that a space is differentiable does not mean that such concepts as metric, length, curvature, geodesics, or parallel transport exist in it. Every differentiable space has smoothness, but without *introduced* additional structure properties it is shapeless and sizeless. So these additional structure properties are possessed by some spaces, but not by all. If, for example, there is to be a measure of distance in a space of N-dimensions, it is something that we must put in for ourselves. That branch of mathematics which adds geodesics, parallel transport, and curvature (shape) to a space is called *affine geometry*. These structures will be studied in the next few sections. We already know that Riemannian geometry adds a metric to the space. It is, as we know, most important to note that Riemannian geometry is built up on the concept of the *distance ds* between two *infinitely proximate points*, rather than on the concept of *finite* distance.

The mathematicians have proved that on every differentiable space (manifold) there exists a Riemannian space. They have also proved that a *Riemannian space is metrically and geodesically complete*. Here *complete* means *always defined*. They have also proved that for any pair of points $u^r(t_1)$ and $u^r(t_2)$ in a Riemannian space there exists a *minimizing* geodesic from $u^r(t_1)$ to $u^r(t_2)$, where t is a parameter.

Note that the metric of a Riemannian space does not determine all the properties of the space. Example: A plane and a cylinder immersed in Euclidean 3-space have the same metric forms for suitable choice of coordinates. These two flat spaces have the same properties "in the small," but they have , obviously, not the same properties "in the large." Example: On a cylinder (but not on the plane) there exist closed circuits which cannot be contracted continuously to a point. It is not possible to put the points of a cylinder into *continuous* one-to-one correspondence with the points of a plane.

Remember that a tensor is a local geometric object. A tensor can interact (addition, multiplication, etc.) with other geometric objects residing at the same place (same event), but it cannot interact with any geometric object at another event. In a *curved* space we cannot regard tensors as arrows linking two events ("point for head and point for tail"). Only the local definition of tensors is wholly viable in curved spaces. Also, in *curved* spaces we cannot move a geometric object from point to point, without carefully specifying how it is to be moved and by what route. However, in *flat* spaces one often does not bother to say where a tensor is located. Students often say: "But what *is* a tensor?" The problem here is one of semantics. It would be clearer not to speak of "tensors in general," but instead speak of given quantities as having or lacking a certain tensor character.

We have seen that the coordinate transformation equations do not explicitly contain such quantities as time and motion. But these equations are extremely general. They are of such a nature that they apply to coordinate systems in any kind of motion. And therefore, also, *tensors, tensor fields, and tensor equations apply to coordinate systems in any kind of motion.*

Nature likes theories that are simple when stated in "coordinate-free" geometric language (stated in tensor form in general coordinates) in, for example, a Riemannian space. So nature must love relativistic theories and dislike Newtonian theories.

44. Partial derivative (ordinary derivative) of a tensor in general N-space

The *partial derivative* of a tensor is denoted by

$$\frac{\partial X^{a\ldots}_{b\ldots}}{\partial u^p} \text{ or } \frac{\partial}{\partial u^p} X^{a\ldots}_{b\ldots} \text{ or } X^{a\ldots}_{b\ldots,p} \text{ or } \partial_p X^{a\ldots}_{b\ldots}$$

and is often referred to as the *ordinary derivative* of a tensor. Partial differentiation of tensors is not tensorial (the partial derivative is not a tensor).

Example: If $\overline{X}^a = \dfrac{\partial \overline{u}^a}{\partial u^b} X^b$, then

$$\frac{\partial \overline{X}^a}{\partial \overline{u}^c} = \frac{\partial}{\partial \overline{u}^c}\left(\frac{\partial \overline{u}^a}{\partial u^b} X^b\right) = \frac{\partial}{\partial u^d}\left(\frac{\partial \overline{u}^a}{\partial u^b} X^b\right)\frac{\partial u^d}{\partial \overline{u}^c}$$

$$= \frac{\partial u^d}{\partial \overline{u}^c}\frac{\partial}{\partial u^d}\left(\frac{\partial \overline{u}^a}{\partial u^b} X^b\right) = \frac{\partial u^d}{\partial \overline{u}^c}\left(\frac{\partial \overline{u}^a}{\partial u^b}\frac{\partial X^b}{\partial u^d} + \frac{\partial^2 \overline{u}^a}{\partial u^d \partial u^b} X^b\right)$$

$$= \frac{\partial \overline{u}^a}{\partial u^b}\frac{\partial u^d}{\partial \overline{u}^c}\frac{\partial X^b}{\partial u^d} + \frac{\partial^2 \overline{u}^a}{\partial u^b \partial u^d}\frac{\partial u^d}{\partial \overline{u}^c} X^b$$

If the first term on the right-hand side *alone* were present, then $\dfrac{\partial X^a}{\partial u^b}$, would be a tensor. Obviously the presence of the second term prevents $\dfrac{\partial X^a}{\partial u^b}$ from being a tensor. However, note that $\dfrac{\partial \Phi}{\partial u^a}$, where $\mathbf{\Phi}$ is an invariant, *is* a covariant tensor of rank one, but that $\dfrac{\partial^2 \Phi}{\partial u^a \partial u^b}$ is *not* a tensor. In this book we shall meet three different types of tensorial differentiation.

45. The Lie derivative of a general tensor field

We wish to differentiate a tensor (or tensor field) in general differentiable space in a tensorial manner. It then turns out that we need to introduce some auxiliary tensor field onto the space. To do so we first need to introduce a space-filling family of curves (a "*congruence*") onto the space. A congruence of curves is defined such that only one curve goes through each point in the (differentiable) space. Given any one curve

$$u^a = u^a(t), \text{ where } t \text{ is a parameter}$$

of the congruence of curves, we define the tangent vector field

$$X^a(u(t)) = \frac{du^a}{dt}$$

along the curve $u^a = u^a(t)$. We do this for every curve in the congruence, and use this to define a tangent vector field

$$X^a(u(t)) = \frac{du^a}{dt}$$

over the whole space. Conversely, a non-zero vector field $X^a(u(t)) = \frac{du^a}{dt}$ defined over the whole space can be used to define a congruence of curves in the space. These curves are called the *trajectories* of $X^a(u(t)) = \frac{du^a}{dt}$, and, of course, the equations of these curves are obtained by solving the ordinary differential equations

$$\frac{du^a}{dt} = X^a(u(t))$$

Assume that the local vector field $X^a(u(t)) = \frac{du^a}{dt}$ and the local congruence of curves resulting from the vector field, has been given. We now wish to differentiate, for example, the tensor field $T^{ab}(u)$ with respect to $X^a(u)$. Let u^a and $\overline{u}^a$ be two different coordinate systems of the space, and let

$$\begin{aligned}\overline{u}^a &= u^a + \delta u^a \\ &= u^a + \delta t \frac{du^a}{dt} \qquad \left(\text{because } \frac{\delta u^a}{\delta t} = \frac{du^a}{dt}\right) \\ &= u^a + \delta t\, X^a\left(u(t)\right)\end{aligned}$$

Then we have

$$\overline{T}^{ab}(\overline{u}) = \frac{\partial \overline{u}^a}{\partial u^c}\frac{\partial \overline{u}^b}{\partial u^d} T^{cd}(u)$$

Here we have $T^{ab}(\overline{u}) = \overline{T}^{ab}(\overline{u})$ when $\delta t = 0$ ($\overline{u}^a = u^a$) (But in general $T^{ab}(\overline{u}) \neq \overline{T}^{ab}(\overline{u})$.) Now we "drag" the tensor T^{ab} at the point with coordinates $\overline{u}^a = u^a$ along the curve passing through this point to a neighboring point with coordinates $\overline{u}^a = u^a + \delta t\, X^a\left(u(t)\right)$. And now, if $\delta t \to 0$ (if $\overline{u}^a \to u^a$), then

$$\overline{T}^{ab}(\overline{u}) \to T^{ab}(\overline{u} = u)$$
$$T^{ab}(\overline{u}) \to T^{ab}(\overline{u} = u)$$

Therefore we define the *Lie derivative* (Lie was a Norwegian mathematician) of T^{ab} with respect to $X^a(u)$, denoted by $L_X T^{ab}$, as

$$L_X T^{ab} = \lim_{\delta t \to 0} \frac{T^{ab}(\overline{u}) - \overline{T}^{ab}(\overline{u})}{\delta t}$$

If follows from the definition that $L_X T^{ab}$ is a contravariant tensor of rank two: At the point of the limit it is the difference of two tensors of the same rank and type, and at this point the two coordinate systems are identical. Remember that δt is an invariant. We are now in a position to prove that

$$L_x T^{ab} = X^c \partial_c T^{ab} - T^{ac} \partial_c X^b - T^{cb} \partial_c X^a$$

Proof

We have $\bar{u}^a = u^a + \delta t\, X^a(u)$. Then

$\dfrac{\partial \bar{u}^a}{\partial u^c} = \dfrac{\partial u^a}{\partial u^c} + \delta t \dfrac{\partial}{\partial u^c} X^a(u) = \delta^a_c + \delta t\, \partial_c X^a(u)$ and

$\dfrac{\partial \bar{u}^b}{\partial u^d} = \dfrac{\partial u^b}{\partial u^d} + \delta t \dfrac{\partial}{\partial u^d} X^b(u) = \delta^b_d + \delta t\, \partial_d X^b(u)$ Thus

$$\bar{T}^{ab}(\bar{u}) = \frac{\partial \bar{u}^a}{\partial u^c} \frac{\partial \bar{u}^b}{\partial u^d} T^{cd}(u)$$

$$= \left[\delta^a_c + \delta t\, \partial_c X^a(u)\right]\left[\delta^b_d + \delta t\, \partial_d X^b(u)\right] T^{cd}(u)$$

$$= \delta^a_c \delta^b_d T^{cd}(u) + \delta t \frac{\partial}{\partial u^d} X^b(u)\, \delta^a_c T^{cd}(u)$$

$$+ \delta t \frac{\partial}{\partial u^c} X^a(u)\, \delta^b_d\, T^{cd}(u) + \delta t \frac{\partial}{\partial u^c} X^a(u)\, \delta t \frac{\partial}{\partial u^d} X^b(u)\, T^{cd}(u)$$

$$= T^{ab}(u) + \left[\frac{\partial}{\partial u^c} X^a(u)\, T^{cb}(u) + \frac{\partial}{\partial u^d} X^b(u)\, T^{ad}(u)\right] \delta t + (\ldots)\left((\delta t)^2\right)$$

$$= T^{ab}(u^c) + \left[\frac{\partial}{\partial u^c} X^a(u)\, T^{cb}(u) + \frac{\partial}{\partial u^c} X^b(u)\, T^{ac}(u)\right] \delta t + (\ldots)\left((\delta t)^2\right)$$

Remember the one-dimensional Taylor's theorem:

$$f(x)=f(a)+f'(a)(x-a)+\frac{f''(a)}{2!}(x-a)^2+\ldots+\frac{f^{(n)}(a)}{n!}(x-a)^n+\frac{f^{(n+1)}(X)}{(n+1)!}(x-a)^{n+1}$$

where X is some number between a and x. Because the N-dimensional Taylor's theorem resembles the one-dimensional formula, we now make the following substitutions:

$$x=\bar{u}^c$$
$$a=u^c$$
$$f(x)=T^{ab}(\bar{u}^c)$$
$$x-a=\bar{u}^c-u^c=\delta t\, X^c(u)$$
$$X=\bar{U}, \text{ where } \bar{U} \text{ is some number between } u^c \text{ and } \bar{u}^c$$

Applying Taylor's theorem, we get

$$T^{ab}(\bar{u})= T^{ab}(\bar{u}^c)=T^{ab}\left(u^c+\delta t\, X^c(u)\right)$$

$$= T^{ab}(u^c)+\left.\frac{\partial}{\partial\bar{u}^c}T^{ab}(\bar{u})\right|_{\bar{u}=u^c}\left(\delta t\, X^c(u)\right)$$

$$+\left.\frac{\frac{\partial^2}{\partial\bar{u}^{c2}}T^{ab}(\bar{u})}{2!}\right|_{\bar{u}=u^c}\left(\delta t\, X^c(u)\right)^2$$

$$+\ \ldots\ +\left.\frac{\frac{\partial^n}{\partial\bar{u}^{c\ n}}T^{ab}(\bar{u})}{n!}\right|_{\bar{u}=u^c}\left(\delta t\, X^c(u)\right)^n$$

$$+\frac{\frac{\partial^{n+1}}{\partial\bar{u}^{c\ n+1}}T^{ab}\left(\bar{U}\right)}{(n+1)!}\left(\delta t\, X^c(u)\right)^{n+1}$$

$$= T^{ab}(u^c)\ +\ \left.\frac{\partial}{\partial\bar{u}^c}T^{ab}(\bar{u})\right|_{\bar{u}=u^c}\left(\delta t\, X^c(u)\right)\ +\ [\ldots\ldots]\left((\delta t)^2\right)$$

Note the summation over c in the terms above: As we know, this is the way tensors and tensor algebra are working. We are now in a position to complete the proof.

$$L_X T^{ab} = \lim_{\delta t \to 0} \frac{T^{ab}(\overline{u}) - \overline{T}^{ab}(\overline{u})}{\delta t}$$

$$= \lim_{\delta t \to 0} \frac{T^{ab}(u^c) + \frac{\partial}{\partial \overline{u}^c} T^{ab}(\overline{u})\Big|_{\overline{u}=u^c} \left(\delta t\, X^c(u)\right) + [......]\left((\delta t)^2\right)}{\delta t} \rightarrow$$

$$\underline{-\left[T^{ab}(u^c) + \left[\frac{\partial}{\partial u^c} X^a(u)\, T^{cb}(u) + \frac{\partial}{\partial u^c} X^b(u)\, T^{ac}(u)\right]\delta t + (...)\left((\delta t)^2\right)\right]}$$

$$= X^c \frac{\partial}{\partial u^c} T^{ab} - T^{ac} \frac{\partial}{\partial u^c} X^b - T^{cb} \frac{\partial}{\partial u^c} X^a$$

$$= X^c \partial_c T^{ab} - T^{ac} \partial_c X^b - T^{cb} \partial_c X^a$$

Of course, the above terms are summed over c. But note also the necessity of summation over c because this is differentiation with respect to (the N components of) X^a. It is not differentiation with respect to u^c.

The Lie derivative of a *scalar field* ϕ is given by

$$L_X \phi = X^a \partial_a \phi$$

The Lie derivative of a *contravariant vector field* Y^a is given by

$$L_X Y^a = X^b \partial_b Y^a - Y^b \partial_b X^a$$

The Lie derivative of a *covariant vector field* is given by

$$L_X Y_a = X^b \partial_b Y_a + Y_b \partial_a X^b$$

The Lie derivative of a general tensor field $T^{a\ldots}_{b\ldots}$ with respect to X^a has the following basic properties:

1. It is linear. Example: $L_X(\lambda Y^a + \mu Z^a) = \lambda L_X Y^a + \mu L_X Z^a$ where λ and μ are constants.
2. It satisfies the usual product rule. Example: $L_X(Y^a Z_{bc}) = Y^a(L_X Z_{bc}) + (L_X Y^a) Z_{bc}$.
3. It is type-preserving. (The Lie derivative of a tensor is again a tensor of the same type.)
4. It commutes with contraction. Example. $\delta^a_b L_X T^{a\Box}_{\Box b} = L_X T^{a\Box}_{\Box a}$.

It can be shown that it is always possible to introduce a coordinate system (in a general differentiable space) such that the curve passing through a point is given by u^1 varying, with $u^2, u^3, \ldots, u^N$ all constant along the curve, and such that

$$X^a = (1, 0, 0, \ldots, 0) = \delta^a_1$$

along this curve in this particular coordinate system. Then, it is easy to see that the equation

$$L_x T^{ab} = X^c \partial_c T^{ab} - T^{ac} \partial_c X^b - T^{cb} \partial_c X^a$$

reduces to

$$L_X T^{ab} = \partial_1 T^{ab}$$

In this particular coordinate system, Lie differentiation reduces to ordinary differentiation.

46. The affine connection and covariant differentiation

Our general N space has N dimensions and is differentiable, but is otherwise devoid of special characteristics. Nonetheless we have already established the algebra of tensors. But additional features must be built into the space before it can be suitable for tensor analysis.

Suppose that the covariant vector X_a is displaced from the point P, at which it is defined, to the neighboring point Q, *without change in magnitude or direction.* This phrase in italics has, of course, no precise meaning as yet. However, in the special case of Euclidean N-space with *homogenous* Cartesian coordinates (the coordinates x^r are called *homogenous* if, and only if, if $ds^2 = dx^k dx^k$), this phrase is interpreted as requiring that the displaced vector shall possess the same components as the original vector. Even in this kind of space, if curvilinear coordinates are being used, the directions of the coordinate axes at the point Q will, in general, be different from their directions at P, and so the components of the displaced vector will *not* be unchanged. Therefore, in a differentiable N-space, the components of the displaced vector will be denoted by $X_a + \delta X_a$. The vector $X_a + \delta X_a$ at Q can now be compared with the field vector $X_a + dX_a$ at Q. So we have

$$(X_a - dX_a) - (X_a + \delta X_a) = dX_a - \delta X_a$$

Clearly $dX_a - \delta X_a$ is a vector (covariant tensor of rank one) at Q. We know that

$$dX_a = \frac{\partial X_a}{\partial x^b} dx^b = X_{a,b}\, dx^b$$

If dX_a were a tensor, then $\dfrac{\partial X_a}{\partial x^b}$ would, by the quotient theorem, be a tensor.

But $\dfrac{\partial X_a}{\partial x^b}$ is not a tensor, so dX_a cannot be a tensor. Let us write

$$dX_a - \delta X_a = X_{a;b}\, dx^b$$

where the quantity $X_{a;b}$ will, by the quotient theorem, be a covariant tensor of rank two. $X_{a;b}$ will be termed the *covariant derivative of* X_a. But now we have to define (infinitesimal) *parallel displacement* of a vector. We are at *liberty* to define the parallel displacement of X_a from P to Q in any way we shall find convenient, but *it is necessary that the definition we adopt shall be in conformity with that adopted in Euclidean N-space which is a special case of the differentiable general N-space.*

Suppose that our differentiable N-space is Euclidean N-space with homogenous Cartesian coordinates y^a and general curvilinear coordinates x^a. And let Y_a be the components of the vector field X_a with respect to the coordinates y^a. Then we have

$$X_a = \frac{\partial y^b}{\partial x^a} Y_b$$

$$Y_a = \frac{\partial x^b}{\partial y^a} X_b$$

If the parallel displacement of the vector X_a to the point Q is carried out, its components Y_a will not change, so $\delta Y_a = 0$. Hence, from calculus

$$\delta X_a = \delta\left(\frac{\partial y^b}{\partial x^a} Y_b\right) = \delta\left(\frac{\partial y^b}{\partial x^a}\right) Y_b + \frac{\partial y^b}{\partial x^a}\delta Y_b$$

$$= \delta\left(\frac{\partial y^b}{\partial x^a}\right) Y_b = \frac{\partial\left(\frac{\partial y^b}{\partial x^a}\right)}{\partial x^c} dx^c\ Y_b = \frac{\partial^2 y^b}{\partial x^a \partial x^c} dx^c\ Y_b$$

$$= \frac{\partial^2 y^b}{\partial x^a \partial x^c} dx^c \frac{\partial x^d}{\partial y^b} X_d = \frac{\partial^2 y^b}{\partial x^a \partial x^c}\frac{\partial x^d}{\partial y^b} X_d\ dx^c$$

$$= \Gamma^d_{ac}\ X_d\ dx^c$$

where $\Gamma^d_{ac} = \frac{\partial^2 y^b}{\partial x^a \partial x^c} \frac{\partial x^d}{\partial y^b}$. So in Euclidean N-space, the δX_a are bilinear forms in the X_d and the dx^c. Therefore, in general differentiable N-space, we shall *define* the δX_a by the equation

$$\delta X_a = \Gamma^d_{ac} X_d \, dx^c$$

determining the N^3 quantities Γ^d_{ac} *arbitrarily* at every point in the space. Of course, we require that the Γ^d_{ac} are continuous functions of the x^a and possesses continuous partial derivatives to the necessary order. This set of quantities Γ^d_{ac} is called an *affinity*. Γ^d_{ac} is not a tensor. An affinity specifies an *affine connection* between the points of the space. Notice that, when defining an affine connection, the choice of the components of the affinity is arbitrary within the selected coordinates. When the coordinate system and the affinity have been chosen, the components of the affinity with respect to any other coordinate system are completely fixed by a transformation law. We can write

$$\begin{aligned} dX_a - \delta X_a &= \frac{\partial X_a}{\partial x^c} dx^c - \Gamma^d_{ac} X_d \, dx^c \\ &= \left(\frac{\partial X_a}{\partial x^c} - \Gamma^d_{ac} X_d \right) dx^c \end{aligned}$$

By the quotient theorem, $\frac{\partial X_a}{\partial x^c} - \Gamma^d_{ac} X_d$ is a covariant tensor of rank two. Thus

$$X_{a;c} \, dx^c = \left(\frac{\partial X_a}{\partial x^c} - \Gamma^d_{ac} X_d \right) dx^c$$

$$\text{or } X_{a;c} = \frac{\partial X_a}{\partial x^c} - \Gamma^d_{ac} X_d$$

$$\text{or } X_{a;b} = \frac{\partial X_a}{\partial x^b} - \Gamma^c_{ab} X_c$$

As we have already said, $X_{a;b} = \frac{\partial X_a}{\partial x^b} - \Gamma^c_{ab} X_c$ is the covariant derivative of X_a We will denote the covariant derivative of X_a by $X_{a;b}$ or $\nabla_b X_a$. Thus we can write

$$\nabla_b X_a = \partial_b X_a - \Gamma^c_{ab} X_c$$

In fact, a differentiable *N*-space which is affinely connected (an affine *N*-space) can function as a suitable space for tensor analysis.
The transformation law for an affinity is

$$\overline{\Gamma}^l_{ij} = \frac{\partial \overline{x}^l}{\partial x^r} \frac{\partial x^s}{\partial \overline{x}^i} \frac{\partial x^t}{\partial \overline{x}^j} \Gamma^r_{st} + \frac{\partial \overline{x}^l}{\partial x^r} \frac{\partial^2 x^r}{\partial \overline{x}^i \partial \overline{x}^j}$$

or, equivalently,

$$\overline{\Gamma}^i_{jk} = \frac{\partial \overline{x}^i}{\partial x^r} \frac{\partial x^s}{\partial \overline{x}^j} \frac{\partial x^t}{\partial \overline{x}^k} \Gamma^r_{st} - \frac{\partial^2 \overline{x}^i}{\partial x^r \partial x^s} \frac{\partial x^r}{\partial \overline{x}^j} \frac{\partial x^s}{\partial \overline{x}^k}$$

We now proceed to prove the above transformation laws for affinities. We have

$$\overline{A}_{i;\,j} = \frac{\partial \overline{A}_i}{\partial \overline{x}^j} - \overline{\Gamma}^k_{ij} \overline{A}_k$$

$$\overline{A}_i = \frac{\partial x^r}{\partial \overline{x}^i} A_r$$

$$\overline{A}_{i;\,j} = \frac{\partial x^s}{\partial \overline{x}^i} \frac{\partial x^t}{\partial \overline{x}^j} A_{s;\,t}$$

and $$\frac{\partial \overline{A}_i}{\partial \overline{x}^j} = \frac{\partial x^r}{\partial \overline{x}^i} \frac{\partial x^l}{\partial \overline{x}^j} \frac{\partial A_r}{\partial x^l} + \frac{\partial^2 x^r}{\partial \overline{x}^i \partial \overline{x}^j} A_r$$

Then

$$\frac{\partial x^s}{\partial \overline{x}^i}\frac{\partial x^t}{\partial \overline{x}^j}A_{s;t} = \frac{\partial x^r}{\partial \overline{x}^i}\frac{\partial x^u}{\partial \overline{x}^j}\frac{\partial A_r}{\partial x^u} + \frac{\partial^2 x^r}{\partial \overline{x}^i \partial \overline{x}^j}A_r - \overline{\Gamma}^k_{ij}\frac{\partial x^r}{\partial \overline{x}^k}A_r$$

But we can substitute for $A_{s;t} = \frac{\partial A_s}{\partial x^t} - \Gamma^r_{st}A_r$ and cancel a pair of identical terms from the two sides of the equation, so that we obtain

$$\frac{\partial x^s}{\partial \overline{x}^i}\frac{\partial x^t}{\partial \overline{x}^j}\left(\frac{\partial A_s}{\partial x^t} - \Gamma^r_{st}A_r\right) = \frac{\partial x^r}{\partial \overline{x}^i}\frac{\partial x^u}{\partial \overline{x}^j}\frac{\partial A_r}{\partial x^u} + \frac{\partial^2 x^r}{\partial \overline{x}^i \partial \overline{x}^j}A_r - \overline{\Gamma}^k_{ij}\frac{\partial x^r}{\partial \overline{x}^k}A_r$$

$$\text{or} \quad \frac{\partial x^s}{\partial \overline{x}^i}\frac{\partial x^t}{\partial \overline{x}^j}\frac{\partial A_s}{\partial x^t} - \frac{\partial x^s}{\partial \overline{x}^i}\frac{\partial x^t}{\partial \overline{x}^j}\Gamma^r_{st}A_r = \frac{\partial x^r}{\partial \overline{x}^i}\frac{\partial x^u}{\partial \overline{x}^j}\frac{\partial A_r}{\partial x^u} + \frac{\partial^2 x^r}{\partial \overline{x}^i \partial \overline{x}^j}A_r - \Gamma^k_{ij}\frac{\partial x^r}{\partial \overline{x}^k}A_r$$

$$\text{or} \quad -\frac{\partial x^s}{\partial \overline{x}^i}\frac{\partial x^t}{\partial \overline{x}^j}\Gamma^r_{st}A_r = \frac{\partial^2 x^r}{\partial \overline{x}^i \partial \overline{x}^j}A_r - \overline{\Gamma}^k_{ij}\frac{\partial x^r}{\partial \overline{x}^k}A_r$$

A_r is an arbitrary vector, so

$$\overline{\Gamma}^k_{ij}\frac{\partial x^r}{\partial \overline{x}^k} = \frac{\partial x^s}{\partial \overline{x}^i}\frac{\partial x^t}{\partial \overline{x}^j}\Gamma^r_{st} + \frac{\partial^2 x^r}{\partial \overline{x}^i \partial \overline{x}^j}$$

Recall that $\delta^l_k = \frac{\partial \overline{x}^l}{\partial \overline{x}^k} = \frac{\partial \overline{x}^l}{\partial x^r}\frac{\partial x^r}{\partial \overline{x}^k}$. Now we multiply both sides of the equation by $\frac{\partial \overline{x}^l}{\partial x^r}$:

$$\overline{\Gamma}^k_{ij}\frac{\partial \overline{x}^l}{\partial x^r}\frac{\partial x^r}{\partial \overline{x}^k} = \frac{\partial \overline{x}^l}{\partial x^r}\frac{\partial x^s}{\partial \overline{x}^i}\frac{\partial x^t}{\partial \overline{x}^j}\Gamma^r_{st} + \frac{\partial \overline{x}^l}{\partial x^r}\frac{\partial^2 x^r}{\partial \overline{x}^i \partial \overline{x}^j}$$

$$\text{or} \quad \overline{\Gamma}^l_{ij} = \frac{\partial \overline{x}^l}{\partial x^r}\frac{\partial x^s}{\partial \overline{x}^i}\frac{\partial x^t}{\partial \overline{x}^j}\Gamma^r_{st} + \frac{\partial \overline{x}^l}{\partial x^r}\frac{\partial^2 x^r}{\partial \overline{x}^i \partial \overline{x}^j}$$

Were it not for the presence of the second term on the right-hand side of this equation Γ^l_{ij} would transform as a tensor. The transformation law is *linear* *inhomogeneous* in the components of an affinity, but a tensor transformation law is *linear homogeneous.* Thus, if all the components of an affinity are zero relative to one coordinate system, they are not necessarily zero relative to another coordinate system. Suppose Γ^k_{ij} and $\Gamma^{k\,*}_{ij}$ are two affinities defined over a region of a *N*-space. Then, clearly,

$$\overline{\Gamma}^k_{ij} - \overline{\Gamma}^{k\,*}_{ij} = \frac{\partial \overline{x}^k}{\partial x^r}\frac{\partial x^s}{\partial \overline{x}^i}\frac{\partial x^t}{\partial \overline{x}^j}\left(\Gamma^r_{st} - \Gamma^{r\,*}_{st}\right)$$

so that *the difference of two affinities is a tensor.* (Of course, the difference is not an affinity.) But the sum of two affinities is neither a tensor nor an affinity. However, *the sum of an affinity* Γ^k_{ij} *and a tensor* A^k_{ij} *is an affinity.*

If $\Gamma^k_{ji} = \Gamma^k_{ij}$, then $\overline{\Gamma}^k_{ji} = \overline{\Gamma}^k_{ij}$. Proof:

$$\begin{aligned}\overline{\Gamma}^k_{ji} &= \frac{\partial \overline{x}^k}{\partial x^r}\frac{\partial x^s}{\partial \overline{x}^j}\frac{\partial x^t}{\partial \overline{x}^i}\Gamma^r_{st} + \frac{\partial \overline{x}^k}{\partial x^r}\frac{\partial^2 x^r}{\partial \overline{x}^j \partial \overline{x}^i}\\ &= \frac{\partial \overline{x}^k}{\partial x^r}\frac{\partial x^t}{\partial \overline{x}^i}\frac{\partial x^s}{\partial \overline{x}^j}\Gamma^r_{ts} + \frac{\partial \overline{x}^k}{\partial x^r}\frac{\partial^2 x^r}{\partial \overline{x}^i \partial \overline{x}^j}\\ &= \overline{\Gamma}^k_{ij}\end{aligned}$$

If Γ^a_{bc} is an affinity, the *torsion* is defined by $T^a_{bc} = \frac{1}{2}\left(\Gamma^a_{bc} - \Gamma^a_{cb}\right)$. It is shown that T^a_{bc} is a tensor. It is called the *torsion tensor.*

If $T^a_{bc} \equiv 0$, then the affinity is *symmetric,* i.e. $\Gamma^a_{bc} = \Gamma^a_{cb}$, and if the affinity is symmetric, then $T^a_{bc} \equiv 0$.

There will, in general, be no coordinate system in which the components of an affinity vanish over a region of an affine *N*-space. But if the affinity is symmetric, then it is always possible to find a coordinate system in which the components all vanish along a given curve. (We omit the proof.) From now on we shall restrict ourselves to symmetric affinities.

We can extend the process of covariant differentiation to tensors of all ranks and types. When an invariant field ϕ suffers parallel displacement from P to Q, its value *will be taken to be* unaltered, so that $\delta\phi = 0$ in all coordinate systems. Hence

$$d\phi - \delta\phi = \frac{\partial\phi}{\partial x^i}dx^i$$

is the counterpart for an invariant of the equation

$$dX_i - \delta X_i = X_{i;j}\, dx^j$$

It follows that

$$\phi_{;i} = \frac{\partial\phi}{\partial x^i}$$

$\phi_{;i}$ is, by the quotient theorem, a covariant tensor of rank one (a covariant vector).

Let X_i be an arbitrary covariant vector, and let Y^i be a contravariant vector field. The value of the invariant X_iY^i, when parallel displaced from P to Q, remains unchanged. Hence

$$\delta\left(X_iY^i\right) = 0 \quad \text{or} \quad (\delta X_i)Y^i + X_i\delta Y^i = 0$$

Using the result

$$\delta X_i = \Gamma^l_{ik}X_l dx^k$$

it follows that

$$\Gamma^l_{ik} X_l dx^k Y^i + X_i \delta Y^i = 0$$

$$\text{or } X_i \delta Y^i = -\Gamma^l_{ik} X_l dx^k Y^i$$

$$\text{or } X_k \delta Y^k = -\Gamma^k_{ij} X_k dx^j Y^i$$

or, since the X_k are arbitrary, $\delta Y^k = -\Gamma^k_{ij} Y^i dx^j$.

It follows that

$$dY^k - \delta Y^k = \frac{\partial Y^k}{\partial x^j} dx^j - \left(-\Gamma^k_{ij} Y^i dx^j\right) = \left(\frac{\partial Y^k}{\partial x^j} + \Gamma^k_{ij} Y^i\right) dx^j$$

Since dx^j is an arbitrary vector and $dY^k - \delta Y^k$ is a vector, then

$$Y^k{}_{;j} = \frac{\partial Y^k}{\partial x^j} + \Gamma^k_{ij} Y^i \text{ or } X^a{}_{;b} = \frac{\partial X^a}{\partial x^b} + \Gamma^a_{cb} X^c$$

is a mixed tensor of rank two. This equation *defines* the covariant derivative of the vector Y^k.

Let Y_i and Z^j be arbitrary vectors, and let X^i_j be a tensor field. The value of the invariant $X^i_j Y_i Z^j$, when parallel displaced from P to Q, remains unchanged. Hence

$$\delta\left(X^i_j Y_i Z^j\right) = 0 \text{ or, from calculus, } \delta X^i_j Y_i Z^j + X^i_j \delta Y_i Z^j + X^i_j Y_i \delta Z^j = 0$$

Using the results

$$\delta Y_i = \Gamma^l_{ik} Y_l dx^k$$

$$\delta Z^j = -\Gamma^j_{ik} Z^i dx^k$$

it follows that

$$\delta X^i_j Y_i Z^j + X^i_j \Gamma^l_{ik} Y_l dx^k Z^j + X^i_j Y_i \left(-\Gamma^j_{ik} Z^i dx^k\right) = 0$$

$$\text{or } \delta X^i_j Y_i Z^j = \Gamma^j_{ik} X^i_j dx^k Y_i Z^i - \Gamma^l_{ik} X^i_j dx^k Y_l Z^j$$

$$\text{or } \delta X^i_j Y_i Z^j = \Gamma^l_{jk} X^i_l dx^k Y_i Z^j - \Gamma^i_{lk} X^l_j dx^k Y_i Z^j$$

or, since the $Y_i Z^j$ are arbitrary,

$$\delta X^i_j = \Gamma^l_{jk} X^i_l dx^k - \Gamma^i_{lk} X^l_j dx^k$$

It follows that

$$dX^i_j - \delta X^i_j = \frac{\partial X^i_j}{\partial x^k} dx^k - \left(\Gamma^l_{jk} X^i_l dx^k - \Gamma^i_{lk} X^l_j dx^k\right) = \left(\frac{\partial X^i_j}{\partial x^k} - \Gamma^l_{jk} X^i_l + \Gamma^i_{lk} X^l_j\right) dx^k$$

Since dx^k is an arbitrary vector and $dX^i_j - \delta X^i_j$ is a mixed tensor of rank two,

$$X^i_{j;k} = \frac{\partial X^i_j}{\partial x^k} - \Gamma^l_{jk} X^i_l + \Gamma^i_{lk} X^l_j$$

is a mixed tensor of rank three. This equation *defines* the covariant derivative of the tensor X^i_j.

The formula for the covariant derivative of a general tensor is

$$X^{r_1 \ldots r_m}_{s_1 \ldots s_n;p} = \frac{\partial}{\partial x^p} X^{r_1 \ldots r_m}_{s_1 \ldots s_n} + \Gamma^{r_1}_{qp} X^{q r_2 \ldots r_m}_{s_1 \ldots s_n} + \ldots + \Gamma^{r_m}_{qp} X^{r_1 \ldots r_{m-1} q}_{s_1 \ldots s_n}$$

$$- \Gamma^q_{s_1 p} X^{r_1 \ldots r_m}_{q s_2 \ldots s_n} - \ldots - \Gamma^q_{s_n p} X^{r_1 \ldots r_m}_{s_1 \ldots s_{n-1} q}$$

$$\text{or } \nabla_c X^{a\ldots}_{b\ldots} = \partial_c X^{a\ldots}_{b\ldots} + \Gamma^a_{dc} X^{d\ldots}_{b\ldots} + \ldots - \Gamma^d_{bc} X^{a\ldots}_{d\ldots} - \ldots$$

Here is another useful result: If the affinity is symmetric, then, in the formula for a Lie derivative of a tensor, *all* occurrences of the partial derivatives may be replaced by covariant derivatives.

The ordinary rules for the differentiation of sums and products apply to the process of covariant differentiation. Examples:

$$\text{If } C^i_j = A^i_j \pm B^i_j \text{, then } C^i_{j;k} = A^i_{j;k} \pm B^i_{j;k}$$

$$\text{If } C^i = A^i_j B^j \text{, then } C^i{}_{;k} = A^i_{j;k} B^j + B^j{}_{;k} A^i_j$$

$$\left(A^i B_j\right)_{;k} = A^i{}_{;k} B_j + B_{j;k} A^i$$

$$\left(A^i B_i\right)_{;k} = \frac{\partial}{\partial x^k}\left(A^i B_i\right) = \frac{\partial A^i}{\partial x^k} B_i + A^i \frac{\partial B_i}{\partial x^k}$$

$$\left(A^{rs} B_{rm}\right)_{;t} = A^{rs}{}_{;t} B_{rm} + A^{rs} B_{rm;t}$$

In covariant differentiation δ^i_j can be treated as a constant because $\delta^i_{j;k} = 0$.

47. The intrinsic (absolute) derivative and affine geodesics

In section 45 we saw that a vector field

$$X^a\left(x(u)\right) = \frac{dx^a}{du}$$

determines a local congruence of curves. The equation for any one curve C of the congruence is

$$x^a = x^a(u)$$

We define the *intrinsic (absolute) derivative* of a tensor $T^{a...}_{b...}$ along a curve C of the congruence, written $\dfrac{\delta T^{a...}_{b...}}{\delta u}$ or $\nabla_X T^{a...}_{b...}$ by the tensor

$$\frac{\delta}{\delta u}\left(T^{a...}_{b...}\right) = T^{a...}_{b...;c}\frac{dx^c}{du} \quad \text{or} \quad \nabla_X T^{a...}_{b...} = X^c \nabla_c T^{a...}_{b...}$$

The tensor $T^{a...}_{b...}$ is said to be *propagated parallely* along the curve C if $\dfrac{\delta}{\delta u}\left(T^{a...}_{b...}\right) = 0$.

This is a first-order ordinary differential equation for $T^{a...}_{b...}$. Given an initial value $T^{a...}_{b...}(P)$, this equation determines a tensor along C which is everywhere parallel to $T^{a...}_{b...}(P)$. An *affine geodesic* is defined as a curve along which the tangent vector is propagated parallel to itself. The parallely propagated vector is, at any point of the curve proportional to the tangent vector at that point. Therefore, the equation for an affine geodesic is

$$\frac{\delta}{\delta u}\left(\frac{dx^a}{du}\right) = \lambda(u)\frac{dx^a}{du} \quad \text{or} \quad \nabla_X X^a = \lambda X^a \text{ or, equivalently,}$$

$$\frac{d^2x^a}{du^2}+\Gamma^a_{bc}\frac{dx^b}{du}\frac{dx^c}{du}=\lambda(u)\frac{dx^a}{du}$$

Proof:

$$\begin{aligned}
\frac{\delta}{\delta u}\left(\frac{dx^a}{du}\right) &= \left(\frac{dx^a}{du}\right)_{;b}\frac{dx^b}{du} \\
&= \left(\frac{\partial\left(\frac{dx^a}{du}\right)}{\partial x^b}+\Gamma^a_{bc}\frac{dx^c}{du}\right)\frac{dx^b}{du} \\
&= \frac{\partial\left(\frac{dx^a}{du}\right)}{\partial x^b}\frac{dx^b}{du}+\Gamma^a_{bc}\frac{dx^b}{du}\frac{dx^c}{du} \\
&= \frac{d^2x^a}{du^2}+\Gamma^a_{bc}\frac{dx^b}{du}\frac{dx^c}{du}
\end{aligned}$$

If the curve C is parametrized in such a way that λ vanishes, then the parameter is a privileged parameter called an *affine parameter*, often denoted by s, and the affine geodesic equation reduces to

$$\frac{d^2x^a}{ds^2}+\Gamma^a_{bc}\frac{dx^b}{ds}\frac{dx^c}{ds}=0.$$

It can be shown that an affine parameter s is only defined up to an *affine transformation*

$s\rightarrow\alpha s+\beta$, where α and β are constants. The *affine length* of the geodesic between two points P_1 and P_2 is defined by

$$\int_{P_1}^{P_2} ds$$

Note that without a metric we cannot, in general, compare lengths on *different* geodesics.

It is evident from the *existence and uniqueness theorem for ordinary differential equations* that corresponding to every direction at a point in a general differentiable N-space there is a unique geodesic passing through the point. It can be proved that any point can be joined to any other point, as long as the points are sufficiently close, by a unique geodesic, but that in the large different geodesics may *refocus* i.e. meet again.

48. The Riemann-Christoffel tensor

As we know already, partial differentiation is in general *commutative* i.e.

$$\frac{\partial^2}{\partial x^c \partial x^d}\left(T^{a\ldots}_{b\ldots}\right) = \frac{\partial^2}{\partial x^d \partial x^c}\left(T^{a\ldots}_{b\ldots}\right)$$

However, covariant differentiation is not in general commutative. The *commutator* of any tensor $T^{a\ldots}_{b\ldots}$ is defined by

$$\left(T^{a\ldots}_{b\ldots;d}\right)_{;c} - \left(T^{a\ldots}_{b\ldots;c}\right)_{;d}$$

so the commutator of a vector A_p is

$$\left(A_{p;q}\right)_{;r} - \left(A_{p;r}\right)_{;q}$$

Let us use the notation

$$A_{p;qr} - A_{p;rq} = \left(A_{p;q}\right)_{;r} - \left(A_{p;r}\right)_{;q}$$

We shall now prove that

$$A_{p;qr} - A_{p;rq} = R^n_{pqr} A_n$$

where A_p is an arbitrary tensor and R^n_{pqr} is a tensor by the quotient law. R^n_{pqr} is called the *Riemann-Christoffel tensor.*

We have

$$A_{p;qr} = \frac{\partial A_{p;q}}{\partial x^r} - \Gamma^j_{pr} A_{j;q} - \Gamma^j_{qr} A_{p;j}$$

$$= \frac{\partial}{\partial x^r}\left(\frac{\partial A_p}{\partial x^q} - \Gamma^j_{pq} A_j\right) - \Gamma^j_{pr}\left(\frac{\partial A_j}{\partial x^q} - \Gamma^k_{jq} A_k\right) - \Gamma^j_{qr}\left(\frac{\partial A_p}{\partial x^j} - \Gamma^l_{pj} A_l\right)$$

$$= \frac{\partial^2 A_p}{dx^r \partial x^q} - \frac{\partial}{\partial x^r}\Gamma^j_{pq} A_j - \Gamma^j_{pq}\frac{\partial A_j}{\partial x^r} - \Gamma^j_{pr}\frac{\partial A_j}{\partial x^q} + \Gamma^j_{pr}\Gamma^k_{jq} A_k - \Gamma^j_{qr}\frac{\partial A_p}{\partial x^j}$$

$$+ \Gamma^j_{qr}\Gamma^l_{pj} A_l$$

and

$$A_{p;rq} = \frac{\partial A_{p;r}}{\partial x^q} - \Gamma^j_{pq} A_{j;r} - \Gamma^j_{rq} A_{p;j}$$

$$= \frac{\partial}{\partial x^q}\left(\frac{\partial A_p}{\partial x^r} - \Gamma^j_{pr} A_j\right) - \Gamma^j_{pq}\left(\frac{\partial A_j}{\partial x^r} - \Gamma^k_{jr} A_k\right) - \Gamma^j_{rq}\left(\frac{\partial A_p}{\partial x^j} - \Gamma^l_{pj} A_l\right)$$

$$= \frac{\partial^2 A_p}{dx^q \partial x^r} - \frac{\partial}{\partial x^q}\Gamma^j_{pr} A_j - \Gamma^j_{pr}\frac{\partial A_j}{\partial x^q} - \Gamma^j_{pq}\frac{\partial A_j}{\partial x^r} + \Gamma^j_{pq}\Gamma^k_{jr} A_k - \Gamma^j_{rq}\frac{\partial A_p}{\partial x^j}$$

$$+ \Gamma^j_{rq}\Gamma^l_{pj} A_l$$

We find

$$A_{p;qr} - A_{p;rq} = \Gamma^j_{pr}\Gamma^k_{jq} A_k - \frac{\partial}{\partial x^r}\Gamma^j_{pq} A_j - \Gamma^j_{pq}\Gamma^k_{jr} A_k + \frac{\partial}{\partial x^q}\Gamma^j_{pr} A_j$$

$$= \Gamma^k_{pr}\Gamma^j_{kq} A_j - \frac{\partial}{\partial x^r}\Gamma^j_{pq} A_j - \Gamma^k_{pq}\Gamma^j_{kr} A_j + \frac{\partial}{\partial x^q}\Gamma^j_{pr} A_j$$

$$= R^j_{pqr} A_j$$

where

$$R^j_{pqr} = \Gamma^k_{pr}\Gamma^j_{kq} - \frac{\partial}{\partial x^r}\Gamma^j_{pq} - \Gamma^k_{pq}\Gamma^j_{kr} + \frac{\partial}{\partial x^q}\Gamma^j_{pr}$$

By replacing j by n, we get the desired result:

$$A_{p;qr} - A_{p;rq} = R^{n}_{pqr} A_{n} .$$

It can be proved that, for a symmetric affinity, the commutator of any tensor can be expressed in terms of the tensor itself and the Riemann-Christoffel tensor. Therefore, the vanishing of the Riemann-Christoffel tensor is a necessary and sufficient condition for the vanishing of the commutator of any tensor.

49. Affine flatness

If there exists a special coordinate system in which the (symmetric) affinity vanishes everywhere, then the space is called *affine flat* or *flat.*

A necessary and sufficient condition for a space to be affine flat (or flat) is that the Riemann tensor vanishes, that is

$$\underset{\text{everywhere}}{\Gamma^a_{bc}=0} \quad \Leftrightarrow \quad \underset{\text{everywhere}}{R^a_{bcd}=0}$$

We have

$$R^j_{pqr}=\Gamma^k_{pr}\Gamma^j_{kq}-\frac{\partial}{\partial x^r}\Gamma^j_{pq}-\Gamma^k_{pq}\Gamma^j_{kr}+\frac{\partial}{\partial x^q}\Gamma^j_{pr}$$

Hence

$$\underset{\text{everywhere}}{\Gamma^a_{bc}=0} \quad \Rightarrow \quad \underset{\text{everywhere}}{R^a_{bcd}=0}$$

Inspection of the above equation makes us think that

$$\underset{\text{everywhere}}{R^a_{bcd}=0} \quad \Rightarrow \quad \underset{\text{everywhere}}{\Gamma^a_{bc}=0}$$

and the mathematicians have proved that this is the case.

Let us construct a tensor field $T^{a\ldots}_{b\ldots}(x)$ over a general affine space by parallely propagating $T^{a\ldots}_{b\ldots}(x)$, that is solving the equation $\dfrac{\delta T^{a\ldots}_{b\ldots}}{\delta u}=T^{a\ldots}_{b\ldots\,;c}\dfrac{dx^c}{du}=0$.

Note that since $\dfrac{dx^c}{du}$ is arbitrary, $T^{a\ldots}_{b\ldots;c}=0$. If we can parallely propagate a tensor from one point to any other point, and the resulting tensor is independent of the curve taken, the the affinity is called *integrable.* If we parallely propagate a tensor from one point to another point along two different curves, and obtain two different tensors, then the affinity is not integrable. Note: In a general affine space, the *intuitive* concept of parallel propagation does not hold. It is shown that: *A necessary and sufficient condition for an affine space to be affine flat (flat) is that the affinity is symmetric and integrable.* Also, it is shown that: *A necessary and sufficient condition for an affinity to be integrable is that the Riemann-Christoffel tensor vanishes.*

Example:

The following results are also proved by the mathematicians: In an affine space, if $R^{a}_{bcd} = 0$ then a vector $\left(T_r\right)_A$ assigned at a point A defines a vector field T_r throughout the space by parallel propagation no matter what paths of propagation are used. If B is any second point in the space, then $\int_A^B T_n dx^n$ is independent of the path of integration. Also, of course,

$$\int_{\bar{A}}^{\bar{B}} \bar{T}_n d\bar{x}^n = \int_A^B T_n dx^n .$$

50. Christoffel's symbols

To work effectively in tensor calculus one needs the *Cristoffel's symbols*. The symbol

$$[pq,r] = \frac{1}{2}\left(\frac{\partial g_{pr}}{\partial x^q} + \frac{\partial g_{qr}}{\partial x^p} - \frac{\partial g_{pq}}{\partial x^r} \right)$$

is called the *Christoffel symbol of the first kind*. The symbol

$$\begin{Bmatrix} s \\ pq \end{Bmatrix} = g^{sr}\,[pq,r]$$

is called the *Christoffel symbol of the second kind*. Neither Christoffel symbol is a tensor. The transformation law for $[pq,r]$ is

$$\left[\overline{jk},\overline{m}\right] = \frac{\partial x^p}{\partial \overline{x}^j}\frac{\partial x^q}{\partial \overline{x}^k}\frac{\partial x^r}{\partial \overline{x}^m}[pq,r] + \frac{\partial^2 x^p}{\partial \overline{x}^j \partial \overline{x}^k}\frac{\partial x^q}{\partial \overline{x}^m}g_{pq}$$

Proof:

Since g_{pq} transforms according to $\overline{g}_{jk} = \dfrac{\partial x^p}{\partial \overline{x}^j}\dfrac{\partial x^q}{\partial \overline{x}^k}g_{pq}$

(1) $$\frac{\partial \overline{g}_{jk}}{\partial \overline{x}^m} = \frac{\partial x^p}{\partial \overline{x}^j}\frac{\partial x^q}{\partial \overline{x}^k}\frac{\partial g_{pq}}{\partial x^r}\frac{\partial x^r}{\partial \overline{x}^m} + \frac{\partial x^p}{\partial \overline{x}^j}\frac{\partial^2 x^q}{\partial \overline{x}^m \partial \overline{x}^k} g_{pq} + \frac{\partial^2 x^p}{\partial \overline{x}^m \partial \overline{x}^j}\frac{\partial x^q}{\partial \overline{x}^k} g_{pq}$$

or, by permutation of the indices,

(2) $$\frac{\partial \overline{g}_{km}}{\partial \overline{x}^j} = \frac{\partial x^q}{\partial \overline{x}^k}\frac{\partial x^r}{\partial \overline{x}^m}\frac{\partial g_{qr}}{\partial x^p}\frac{\partial x^p}{\partial \overline{x}^j} + \frac{\partial x^q}{\partial \overline{x}^k}\frac{\partial^2 x^r}{\partial \overline{x}^j \partial \overline{x}^m} g_{qr} + \frac{\partial^2 x^q}{\partial \overline{x}^j \partial \overline{x}^k}\frac{\partial x^r}{\partial \overline{x}^m} g_{qr}$$

(3) $$\frac{\partial \overline{g}_{mj}}{\partial \overline{x}^k} = \frac{\partial x^r}{\partial \overline{x}^m}\frac{\partial x^p}{\partial \overline{x}^j}\frac{\partial g_{rp}}{\partial x^q}\frac{\partial x^q}{\partial \overline{x}^k} + \frac{\partial x^r}{\partial \overline{x}^m}\frac{\partial^2 x^p}{\partial \overline{x}^k \partial \overline{x}^j} g_{rp} + \frac{\partial^2 x^r}{\partial \overline{x}^k \partial \overline{x}^m}\frac{\partial x^p}{\partial \overline{x}^j} g_{rp}$$

Doing the calculation $\frac{1}{2}[(2)+(3)-(1)]$ we obtain on using the definition of the Christoffel symbol of the first kind

$$\left[\overline{jk},\overline{m}\right] = \frac{\partial x^p}{\partial \overline{x}^j}\frac{\partial x^q}{\partial \overline{x}^k}\frac{\partial x^r}{\partial \overline{x}^m}[pq,r] + \frac{\partial^2 x^p}{\partial \overline{x}^j \partial \overline{x}^k}\frac{\partial x^q}{\partial \overline{x}^m} g_{pq}\,.$$

The transformation law for $\begin{Bmatrix} s \\ pq \end{Bmatrix}$ is

$$\begin{Bmatrix} \overline{n} \\ \overline{jk} \end{Bmatrix} = \frac{\partial \overline{x}^n}{\partial x^s}\frac{\partial x^p}{\partial \overline{x}^j}\frac{\partial x^q}{\partial \overline{x}^k}\begin{Bmatrix} s \\ pq \end{Bmatrix} + \frac{\partial \overline{x}^n}{\partial x^s}\frac{\partial^2 x^s}{\partial \overline{x}^j \partial \overline{x}^k}$$

Proof:

Since $\overline{g}^{nm} = \frac{\partial \overline{x}^n}{\partial x^s}\frac{\partial \overline{x}^m}{\partial x^t} g^{st}$, $\frac{\partial x^r}{\partial \overline{x}^m}\frac{\partial \overline{x}^m}{\partial x^t} = \delta^r_t$, and $g^{sq} g_{pq} = \delta^s_p$,

$$\left\{ \begin{matrix} \bar{n} \\ \bar{j}\bar{k} \end{matrix} \right\} = \bar{g}^{nm} \left[\bar{j}\bar{k}, \bar{m} \right]$$

$$= \frac{\partial x^p}{\partial \bar{x}^j} \frac{\partial x^q}{\partial \bar{x}^k} \frac{\partial x^r}{\partial \bar{x}^m} \frac{\partial \bar{x}^n}{\partial x^s} \frac{\partial \bar{x}^m}{\partial x^t} g^{st} [pq, r] + \frac{\partial^2 x^p}{\partial \bar{x}^j \partial \bar{x}^k} \frac{\partial x^q}{\partial \bar{x}^m} \frac{\partial \bar{x}^n}{\partial x^s} \frac{\partial \bar{x}^m}{\partial x^t} g^{st} g_{pq}$$

$$= \frac{\partial x^p}{\partial \bar{x}^j} \frac{\partial x^q}{\partial \bar{x}^k} \frac{\partial \bar{x}^n}{\partial x^s} \delta_t^r g^{st} [pq, r] + \frac{\partial^2 x^p}{\partial \bar{x}^j \partial \bar{x}^k} \frac{\partial \bar{x}^n}{\partial x^s} \delta_t^q g^{st} g_{pq}$$

$$= \frac{\partial x^p}{\partial \bar{x}^j} \frac{\partial x^q}{\partial \bar{x}^k} \frac{\partial \bar{x}^n}{\partial x^s} g^{sr} [pq, r] + \frac{\partial^2 x^p}{\partial \bar{x}^j \partial \bar{x}^k} \frac{\partial \bar{x}^n}{\partial x^s} g^{sq} g_{pq}$$

$$= \frac{\partial x^p}{\partial \bar{x}^j} \frac{\partial x^q}{\partial \bar{x}^k} \frac{\partial \bar{x}^n}{\partial x^s} \left\{ \begin{matrix} s \\ pq \end{matrix} \right\} + \frac{\partial^2 x^p}{\partial \bar{x}^j \partial \bar{x}^k} \frac{\partial \bar{x}^n}{\partial x^s} \delta_p^s$$

$$= \frac{\partial x^p}{\partial \bar{x}^j} \frac{\partial x^q}{\partial \bar{x}^k} \frac{\partial \bar{x}^n}{\partial x^s} \left\{ \begin{matrix} s \\ pq \end{matrix} \right\} + \frac{\partial^2 x^p}{\partial \bar{x}^j \partial \bar{x}^k} \frac{\partial \bar{x}^n}{\partial x^p}$$

$$= \frac{\partial \bar{x}^n}{\partial x^s} \frac{\partial x^p}{\partial \bar{x}^j} \frac{\partial x^q}{\partial \bar{x}^k} \left\{ \begin{matrix} s \\ pq \end{matrix} \right\} + \frac{\partial \bar{x}^n}{\partial x^s} \frac{\partial^2 x^s}{\partial \bar{x}^j \partial \bar{x}^k}$$

The following identities are useful:

$$[pq,r]=[qp,r]$$

$$\begin{Bmatrix} s \\ pq \end{Bmatrix}=\begin{Bmatrix} s \\ qp \end{Bmatrix}$$

$$[pq,r]=g_{rs}\begin{Bmatrix} s \\ pq \end{Bmatrix}$$

$$\frac{\partial g_{pq}}{\partial x^m}=[pm,q]+[qm,p]$$

Proofs:

$$\begin{aligned}[pq,r]&=\frac{1}{2}\left(\frac{\partial g_{pr}}{\partial x^q}+\frac{\partial g_{qr}}{\partial x^p}-\frac{\partial g_{pq}}{\partial x^r}\right)\\ &=\frac{1}{2}\left(\frac{\partial g_{qr}}{\partial x^p}+\frac{\partial g_{pr}}{\partial x^q}-\frac{\partial g_{qp}}{\partial x^r}\right)\\ &=[qp,r]\end{aligned}$$

$$\begin{Bmatrix} s \\ pq \end{Bmatrix}=g^{sr}[pq,r]=g^{sr}[qp,r]=\begin{Bmatrix} s \\ qp \end{Bmatrix}$$

$$g_{ks}\begin{Bmatrix} s \\ pq \end{Bmatrix}=g_{ks}g^{sr}[pq,r]=\delta_k^r[pq,r]=[pq,k]$$

or $[pq,r]=g_{rs}\begin{Bmatrix} s \\ pq \end{Bmatrix}$

$$[pm, q] + [qm, p]$$

$$= \frac{1}{2}\left(\frac{\partial g_{pq}}{\partial x^m} + \frac{\partial g_{mq}}{\partial x^p} - \frac{\partial g_{pm}}{\partial x^q}\right) + \frac{1}{2}\left(\frac{\partial g_{qp}}{\partial x^m} + \frac{\partial g_{mp}}{\partial x^q} - \frac{\partial g_{qm}}{\partial x^p}\right)$$

$$= \frac{\partial g_{pq}}{\partial x^m}$$

51. Metric geodesics

Riemannian geometry is not built up on the concept of finite distance, but on the concept of
ds. By definition *ds* is the length of the displacement dx^r. As we already know (see section 39): The distance *s* between two points $x^r(u_1)$ and $x^r(u_2)$ on a curve $x^r = x^r(u)$ in a general Riemannian *N*-space, with general coordinates x^r, having metric

$$ds^2 = \varepsilon g_{ij} dx^i dx^j, \quad ds \succ 0$$

is given by

$$s = \int_{u_1}^{u_2} \sqrt{\varepsilon g_{ij} \frac{dx^i}{du} \frac{dx^j}{du}} du$$

where $\varepsilon = +1$ or $\varepsilon = -1$, so as to make $ds^2 \succ 0$. ε is called the *indicator* of dx^r.

That curve $x^r = x^r(u)$ which makes the distance *s* a minimum is called a *metric geodesic* of the Riemannian *N*-space. The great circles on a sphere are curves of *stationary* length. They are analogous to the straight lines of Euclidean space. The idea of stationary length and the idea of shortest length are equivalent as a basis of definition of a metric geodesic of a Riemannian *N*-space.

The differential equations of a metric geodesic in a Riemannian *N*-space (whether the metric form is positive-definite or negative-definite) read

$$\frac{d^2x^r}{ds^2} + \begin{Bmatrix} r \\ mn \end{Bmatrix} \frac{dx^m}{ds} \frac{dx^n}{ds} = 0$$

where *s* is the *arc length parameter* and we have assumed that $ds \neq 0$ at any point of the curve. The above equations will be proved later. Of course, if these equations are satisfied along the curve $x^r = x^r(s)$, then the curve is a metric geodesic. These N equations are second-order differential equations for the functions $x^r = x^r(s)$. Their solution involves $2N$ constants of integration. A solution $x^r = x^r(s)$ is, in general, determined uniquely if we are given $2N$ conditions for $x^r = x^r(s)$. Hence, a metric geodesic is, in general, determined uniquely if we are given a point on it and the direction of the tangent at that point: $x^r = a^r$ and $\left.\frac{dx^r}{ds}\right|_{x^r = a^r}$ or if we are given two distinct points $x^i = a^i$ and $x^i = b^i$ on it. There is, in general, a unique metric geodesic connecting every pair of points. But in some cases this will not be so. For example, the metric geodesics of the surface of a sphere (a Riemannian 2-space) are great circles and there are two great circle arcs, or there are an infinity of great circle arcs, joining two given points. Also, it can be shown that it is, in general, possible to draw at least one geodesic through two given points in a general Riemannian N-space. We shall now give a proof of the above differential equations.

The equations $x^r = x^r(u,v)$ represent a singly infinite family of curves if we hold v fixed, and let u vary along each curve. Let a singly infinite family of curves $x^r = x^r(u,v)$ join common *end points* A and B. Suppose the parameter u has the same values u_1 and u_2 for all the curves at A and B, respectively. The length of any curve of this family is

$$L = \int_A^B ds = \int_{u_1}^{u_2} \sqrt{\varepsilon g_{mn} \frac{\partial x^m}{\partial u} \frac{\partial x^n}{\partial u}} du$$

if we suppose that all the curves have the same indicator. A metric geodesic joining points A and B in a general Riemannian N-space satisfies $\delta \int_A^B ds = 0$ by the idea of stationary length.

For shortness we can write

$$g_{mn} \frac{\partial x^m}{\partial u} \frac{\partial x^n}{\partial u} = g_{mn} p^m p^n = \omega \text{ (where } p^m = \frac{\partial x^m}{\partial u}, \quad p^n = \frac{\partial x^n}{\partial u})$$

and

$$L = \int_{u_1}^{u_2} (\varepsilon\omega)^{\frac{1}{2}} du$$

L is a function of v, so (from calculus) for any curve

$$\frac{dL}{dv} = \int_{u_1}^{u_2} \frac{\partial}{\partial v}(\varepsilon\omega)^{\frac{1}{2}}\, du\,.$$

ω is a function of x^r and p^r, and x^r and p^r are functions of u and v. Therefore

$$\frac{\partial}{\partial v}(\varepsilon\omega)^{\frac{1}{2}} = \frac{\partial}{\partial x^r}(\varepsilon\omega)^{\frac{1}{2}} \frac{\partial x^r}{\partial v} + \frac{\partial}{\partial p^r}(\varepsilon\omega)^{\frac{1}{2}} \frac{\partial p^r}{\partial v}\,.$$

But since

$$\frac{\partial p^r}{\partial v} = \frac{\partial}{\partial v}\frac{\partial x^r}{\partial u} = \frac{\partial}{\partial u}\frac{\partial x^r}{\partial v}$$

we see that

$$\frac{\partial}{\partial v}(\varepsilon\omega)^{\frac{1}{2}} = \frac{\partial}{\partial x^r}(\varepsilon\omega)^{\frac{1}{2}} \frac{\partial x^r}{\partial v} + \frac{\partial}{\partial p^r}(\varepsilon\omega)^{\frac{1}{2}} \frac{\partial}{\partial u}\frac{\partial x^r}{\partial v}$$

and that

$$\frac{dL}{dv} = \int_{u_1}^{u_2} \left(\frac{\partial}{\partial x^r}(\varepsilon\omega)^{\frac{1}{2}} \frac{\partial x^r}{\partial v} + \frac{\partial}{\partial p^r}(\varepsilon\omega)^{\frac{1}{2}} \frac{\partial}{\partial u}\frac{\partial x^r}{\partial v} \right) du\,.$$

But

$$\frac{\partial}{\partial u}\left(\frac{\partial}{\partial p^r}(\varepsilon\omega)^{\frac{1}{2}}\frac{\partial x^r}{\partial v}\right)=\frac{\partial}{\partial u}\left(\frac{\partial}{\partial p^r}(\varepsilon\omega)^{\frac{1}{2}}\right)\frac{\partial x^r}{\partial v}+\frac{\partial}{\partial p^r}(\varepsilon\omega)^{\frac{1}{2}}\frac{\partial}{\partial u}\frac{\partial x^r}{\partial v} \quad \text{or}$$

$$\frac{\partial}{\partial p^r}(\varepsilon\omega)^{\frac{1}{2}}\frac{\partial}{\partial u}\frac{\partial x^r}{\partial v}=\frac{\partial}{\partial u}\left(\frac{\partial}{\partial p^r}(\varepsilon\omega)^{\frac{1}{2}}\frac{\partial x^r}{\partial v}\right)-\frac{\partial}{\partial u}\left(\frac{\partial}{\partial p^r}(\varepsilon\omega)^{\frac{1}{2}}\right)\frac{\partial x^r}{\partial v},$$

and so we get

$$\frac{dL}{dv}=\int_{u_1}^{u_2}\left(\frac{\partial}{\partial x^r}(\varepsilon\omega)^{\frac{1}{2}}\frac{\partial x^r}{\partial v}+\frac{\partial}{\partial u}\left(\frac{\partial}{\partial p^r}(\varepsilon\omega)^{\frac{1}{2}}\frac{\partial x^r}{\partial v}\right)-\frac{\partial}{\partial u}\left(\frac{\partial}{\partial p^r}(\varepsilon\omega)^{\frac{1}{2}}\right)\frac{\partial x^r}{\partial v}\right)du$$

$$=\int_{u_1}^{u_2}\frac{\partial}{\partial u}\left(\frac{\partial}{\partial p^r}(\varepsilon\omega)^{\frac{1}{2}}\frac{\partial x^r}{\partial v}\right)du-\int_{u_1}^{u_2}\left(\frac{\partial}{\partial u}\left(\frac{\partial}{\partial p^r}(\varepsilon\omega)^{\frac{1}{2}}\right)\frac{\partial x^r}{\partial v}-\frac{\partial}{\partial x^r}(\varepsilon\omega)^{\frac{1}{2}}\frac{\partial x^r}{\partial v}\right)du$$

$$=\left[\frac{\partial}{\partial p^r}(\varepsilon\omega)^{\frac{1}{2}}\frac{\partial x^r}{\partial v}\right]_{u_1}^{u_2}-\int_{u_1}^{u_2}\left(\frac{\partial}{\partial u}\left(\frac{\partial}{\partial p^r}(\varepsilon\omega)^{\frac{1}{2}}\right)-\frac{\partial}{\partial x^r}(\varepsilon\omega)^{\frac{1}{2}}\right)\frac{\partial x^r}{\partial v}du\,.$$

The change in length δL when we let v vary from v to $v + \delta v$, is

$$\delta L=\frac{dL}{dv}\delta v\,.$$

The increment δx^r in x^r when we pass from a point on the curve v to the point on the curve $v + \delta v$ with the same value of u, is

$$\delta x^r=\frac{\partial x^r}{\partial v}\delta v\,.$$

Therefore, for any curve v,

$$\delta L = \frac{dL}{dv}\delta v = \left[\frac{\partial}{\partial p^r}(\varepsilon\omega)^{\frac{1}{2}}\frac{\partial x^r}{\partial v}\delta v\right]_{u_1}^{u_2} - \int_{u_1}^{u_2}\left(\frac{\partial}{\partial u}\left(\frac{\partial}{\partial p^r}(\varepsilon\omega)^{\frac{1}{2}}\right) - \frac{\partial}{\partial x^r}(\varepsilon\omega)^{\frac{1}{2}}\right)\frac{\partial x^r}{\partial v}\delta v d$$

$$= \left[\frac{\partial}{\partial p^r}(\varepsilon\omega)^{\frac{1}{2}}\delta x^r\right]_{u_1}^{u_2} - \int_{u_1}^{u_2}\left(\frac{\partial}{\partial u}\left(\frac{\partial}{\partial p^r}(\varepsilon\omega)^{\frac{1}{2}}\right) - \frac{\partial}{\partial x^r}(\varepsilon\omega)^{\frac{1}{2}}\right)\delta x^r du$$

$\delta x^r = 0$ at the fixed points A and B, so we have

$$\delta L = -\int_{u_1}^{u_2}\left(\frac{\partial}{\partial u}\left(\frac{\partial}{\partial p^r}(\varepsilon\omega)^{\frac{1}{2}}\right) - \frac{\partial}{\partial x^r}(\varepsilon\omega)^{\frac{1}{2}}\right)\delta x^r du .$$

Therefore, if the particular curve v is a metric geodesic,

$$\int_{u_1}^{u_2}\left(\frac{\partial}{\partial u}\left(\frac{\partial}{\partial p^r}(\varepsilon\omega)^{\frac{1}{2}}\right) - \frac{\partial}{\partial x^r}(\varepsilon\omega)^{\frac{1}{2}}\right)\delta x^r du = 0$$

or, by the *fundamental lemma of the calculus of variations*,

$$\frac{\partial}{\partial u}\left(\frac{\partial}{\partial p^r}(\varepsilon\omega)^{\frac{1}{2}}\right) - \frac{\partial}{\partial x^r}(\varepsilon\omega)^{\frac{1}{2}} = 0 \text{ or}$$

$$\frac{d}{du}\frac{\partial}{\partial p^r}(\varepsilon\omega)^{\frac{1}{2}} - \frac{\partial}{\partial x^r}(\varepsilon\omega)^{\frac{1}{2}} = 0$$

These equations, called *Euler's equations*, are, of course, satisfied at all points on a metric geodesic. They may be written

$$\frac{d}{du}\frac{\partial\omega}{\partial p^r}-\frac{\partial\omega}{\partial x^r}=\frac{1}{2\omega}\frac{d\omega}{du}\frac{\partial\omega}{\partial p^r}$$

Proof

$$\frac{d}{du}\frac{\partial}{\partial p^r}(\varepsilon\omega)^{\frac{1}{2}}-\frac{\partial}{\partial x^r}(\varepsilon\omega)^{\frac{1}{2}}$$

$$=\frac{d}{du}\left(\frac{1}{2}(\varepsilon\omega)^{-\frac{1}{2}}\frac{\partial(\varepsilon\omega)}{\partial p^r}\right)-\frac{1}{2}(\varepsilon\omega)^{-\frac{1}{2}}\frac{\partial(\varepsilon\omega)}{\partial x^r}$$

$$=-\frac{1}{4}(\varepsilon\omega)^{-\frac{3}{2}}\frac{d}{du}(\varepsilon\omega)\frac{\partial(\varepsilon\omega)}{\partial p^r}+\frac{1}{2}(\varepsilon\omega)^{-\frac{1}{2}}\frac{d}{du}\frac{\partial(\varepsilon\omega)}{\partial p^r}-\frac{1}{2}(\varepsilon\omega)^{-\frac{1}{2}}\frac{\partial(\varepsilon\omega)}{\partial x^r}$$

$$=-\frac{1}{2}\frac{1}{2}(\varepsilon\omega)^{-\frac{1}{2}}(\varepsilon\omega)^{-1}\frac{d(\varepsilon\omega)}{du}\frac{\partial(\varepsilon\omega)}{\partial p^r}+\frac{1}{2}(\varepsilon\omega)^{-\frac{1}{2}}\frac{d}{du}\frac{\partial(\varepsilon\omega)}{\partial p^r}-\frac{1}{2}(\varepsilon\omega)^{-\frac{1}{2}}\frac{\partial(\varepsilon\omega)}{\partial x^r}=0$$

or

$$-\frac{1}{2}(\varepsilon\omega)^{-1}\frac{d\omega}{du}\frac{\partial\omega}{\partial p^r}+\frac{d}{du}\frac{\partial(\varepsilon\omega)}{\partial p^r}-\frac{\partial(\varepsilon\omega)}{\partial x^r}=0 \text{ or}$$

$$-\frac{1}{2}\frac{1}{\omega}\frac{d\omega}{du}\frac{\partial\omega}{\partial p^r}+\frac{d}{du}\frac{\partial\omega}{\partial p^r}-\frac{\partial\omega}{\partial x^r}=0 \text{ or}$$

$$\frac{d}{du}\frac{\partial\omega}{\partial p^r}-\frac{\partial\omega}{\partial x^r}=\frac{1}{2\omega}\frac{d\omega}{du}\frac{\partial\omega}{\partial p^r}.$$

If we choose the arbitrary parameter u equal to the arc-length s along the metric geodesic, then

$$u=s,\quad p^r=\frac{dx^r}{ds},\quad \omega=g_{mn}p^m p^n=\varepsilon \quad\text{and}\quad \frac{d\omega}{du}=0,\quad \text{because}$$

$$\varepsilon g_{mn} dx^m dx^n = ds^2 \quad \text{or}$$

$$\varepsilon g_{mn} \frac{dx^m}{ds} \frac{dx^n}{ds} = 1 \quad \text{or}$$

$$g_{mn} \frac{dx^m}{ds} \frac{dx^n}{ds} = \varepsilon$$

Now the differential equations of a metric geodesic read

$$\frac{d}{ds} \frac{\partial \omega}{\partial p^r} - \frac{\partial \omega}{\partial x^r} = 0 .$$

But

$$\frac{\partial \omega}{\partial p^r} = \frac{\partial}{\partial p^r} \left(g_{mn} p^m p^n \right) = g_{mn} p^m \frac{\partial p^n}{\partial p^r} + g_{mn} \frac{\partial p^m}{\partial p^r} p^n + \frac{\partial g_{mn}}{\partial p^r} p^m p^n$$

$$= g_{mr} p^m + g_{rn} p^n + 0 \cdot p^m p^n = 2 g_{rm} p^m ,$$

and

$$\frac{\partial \omega}{\partial x^r} = \frac{\partial}{\partial x^r} \left(g_{mn} p^m p^n \right) = g_{mn} p^m \frac{\partial p^n}{\partial x^r} + g_{mn} \frac{\partial p^m}{\partial x^r} p^n + \frac{\partial g_{mn}}{\partial x^r} p^m p^n$$

$$= 0 + 0 + \frac{\partial g_{mn}}{\partial x^r} p^m p^n = \frac{\partial g_{mn}}{\partial x^r} p^m p^n ,$$

so

$$\frac{d}{ds} \left(2 g_{rm} p^m \right) - \frac{\partial g_{mn}}{\partial x^r} p^m p^n = 0$$

or

$$2\left(\frac{\partial g_{rm}}{\partial x^n}\frac{dx^n}{ds}p^m + g_{rm}\frac{dp^n}{ds}\right) - \frac{\partial g_{mn}}{\partial x^r}p^m p^n = 0$$

or

$$g_{rm}\frac{dp^m}{ds} + \frac{\partial g_{rm}}{\partial x^n}p^m p^n - \frac{1}{2}\frac{\partial g_{mn}}{\partial x^r}p^m p^n = 0\,.$$

Using the identity

$$\frac{\partial g_{rm}}{\partial x^n}p^m p^n = \frac{1}{2}\left(\frac{\partial g_{rm}}{\partial x^n} + \frac{\partial g_{rn}}{\partial x^m}\right)p^m p^n,$$

the equations of a metric geodesic may be written

$$g_{rm}\frac{dp^m}{ds} + \frac{1}{2}\left(\frac{\partial g_{rm}}{\partial x^n} + \frac{\partial g_{rn}}{\partial x^m} - \frac{\partial g_{mn}}{\partial x^r}\right)p^m p^n = 0 \text{ or}$$

$$g_{rm}\frac{dp^m}{ds} + [mn, r]\,p^m p^n = 0 \text{ or}$$

$$g_{sm}\frac{dp^m}{ds} + [mn, s]\,p^m p^n = 0 \text{ or}$$

$$g^{rs}g_{sm}\frac{dp^m}{ds} + g^{rs}[mn, s]\,p^m p^n = 0 \text{ or}$$

$$\delta^r_m\frac{dp^m}{ds} + \begin{Bmatrix} r \\ mn \end{Bmatrix}p^m p^n = 0 \text{ or}$$

$$\frac{dp^r}{ds} + \begin{Bmatrix} r \\ mn \end{Bmatrix}p^m p^n = 0 \text{ or}$$

$$\frac{d^2x^r}{ds^2}+\begin{Bmatrix} r \\ mn \end{Bmatrix}\frac{dx^m}{ds}\frac{dx^n}{ds}=0 .$$

Example:

The geodesic equation for Euclidean 3-space with cylindrical polar coordinates:

In cylindrical polar coordinates $x^1 = R,\ x^2 = \phi,\ x^3 = z$, $ds^2 = dR^2 + R^2 d\phi^2 + dz^2$, $g_{11} = 1$, $g_{22} = R^2$, $g_{33} = 1$ (the only non-zero g_{ab}), $\begin{Bmatrix} 1 \\ 22 \end{Bmatrix} = -R$, $\begin{Bmatrix} 2 \\ 12 \end{Bmatrix} = \frac{1}{R}$, $\begin{Bmatrix} 2 \\ 21 \end{Bmatrix} = \frac{1}{R}$ (the only non-zero $\begin{Bmatrix} a \\ bc \end{Bmatrix}$). The general geodesic equations are $\frac{d^2x^a}{ds^2}+\begin{Bmatrix} a \\ bc \end{Bmatrix}\frac{dx^a}{ds}\frac{dx^b}{ds}=0$, so we have

$$\frac{d^2x^1}{ds^2}+\begin{Bmatrix} 1 \\ 22 \end{Bmatrix}\frac{dx^2}{ds}\frac{dx^2}{ds}=0 \quad \text{or} \quad \frac{d^2R}{ds^2}-R\left(\frac{d\phi}{ds}\right)^2=0$$

$$\frac{d^2x^2}{ds^2}+\begin{Bmatrix} 2 \\ 12 \end{Bmatrix}\frac{dx^1}{ds}\frac{dx^2}{ds}+\begin{Bmatrix} 2 \\ 21 \end{Bmatrix}\frac{dx^2}{ds}\frac{dx^1}{ds}=0 \quad \text{or} \quad \frac{d^2\phi}{ds^2}+\frac{2}{R}\frac{dR}{ds}\frac{d\phi}{ds}=0$$

$$\frac{d^2x^3}{ds^2}=0 \quad \text{or} \quad \frac{d^2z}{ds^2}=0 .$$

52. The metric connection, the covariant derivative, and the absolute derivative

Suppose we have a differentiable N-space endowed with both an affinity and a metric. Compare the equations

$$\frac{d^2x^r}{ds^2} + \Gamma^r_{mn}\frac{dx^m}{ds}\frac{dx^n}{ds} = 0$$

and the equations

$$\frac{d^2x^r}{ds^2} + \begin{Bmatrix} r \\ mn \end{Bmatrix}\frac{dx^m}{ds}\frac{dx^n}{ds} = 0.$$

We have already seen that Γ^r_{mn} and $\begin{Bmatrix} r \\ mn \end{Bmatrix}$ have identical transformation laws, and that both Γ^r_{mn} and $\begin{Bmatrix} r \\ mn \end{Bmatrix}$ are symmetric. Therefore, of course, we can take $\Gamma^r_{mn} = \begin{Bmatrix} r \\ mn \end{Bmatrix}$. The special affinity $\Gamma^r_{mn} = \begin{Bmatrix} r \\ mn \end{Bmatrix}$ is called the *metric connection* (or *metric affinity*). From now on, we shall *always* work with the metric connection, and we shall denote it by $\begin{Bmatrix} r \\ mn \end{Bmatrix}$ rather than Γ^r_{mn}.

Of course, the above results automatically give us the covariant derivative and the intrinsic derivative of a tensor field in a general Riemannian N-space with general coordinates.

It can be proved that in a general Riemannian N-space with general coordinates we have:

$$g_{jk;q} = 0$$

$$g^{jk}{}_{;q} = 0$$

$$\delta^{j}_{k;q} = 0$$

$$\frac{\delta g_{jk}}{\delta t} = 0$$

$$\frac{\delta g^{jk}}{\delta t} = 0$$

$$\frac{\delta \delta^{j}_{k}}{\delta t} = 0$$

53. Metric flatness

The distinction between *flat* and *curved* Riemannian *N*-space is that in a flat Riemannian *N*-space it is possible to find a coordinate system in which the metric form is *everywhere*

$$ds^2 = \pm dx^1 dx^1 \pm dx^2 dx^2 \pm \ldots \pm dx^N dx^N$$

so that the metric components $g_{ij} = \pm 1$ if $i = j$ and zero otherwise. In a curved Riemannian *N*-space one can find a coordinate system in which the metric form is

$$ds^2 = \pm dx^1 dx^1 \pm dx^2 dx^2 \pm \ldots \pm dx^N dx^N$$

at any specified *point* P, but there is no coordinate system giving this metric form *everywhere*. Of course, if a Riemannian *N*-space is flat, then the g_{ab} is (are) constant *everywhere*. Since the g_{ab} is constant, its partial derivatives (ordinary derivatives) vanish, and so the metric connection vanishes everywhere because

$$\begin{Bmatrix} a \\ bc \end{Bmatrix} = g^{ad}\left[bc, d\right] = \frac{1}{2} g^{ad} \left(\frac{\partial g_{bd}}{\partial x^c} + \frac{\partial g_{cd}}{\partial x^b} - \frac{\partial g_{bc}}{\partial x^d} \right).$$

Since the metric connection vanishes everywhere, then so must its partial derivatives (ordinary derivatives). And therefore the Riemann tensor vanishes. It now follows that a *necessary* condition for a Riemannian *N*-space to be flat is that its Riemann tensor vanishes. Conversely, it can be shown that if the Riemann tensor vanishes, then there exists a special coordinate system in which the metric connection vanishes *everywhere* and the g_{ab} are constant *everywhere*, and g_{ab} can be transformed into ± 1. It therefore follows that a *sufficient* condition for a Riemannian *N*-space to be flat is that the Riemann tensor vanishes. Putting the above results together we get the following theorem: *A necessary and sufficient condition for a Riemannian N-space to be flat is that the Riemann tensor vanishes.* Combining this theorem with the theorems of sections 49. and 52., we see that *if we use the metric connection, then metric flatness coincides with affine flatness.*

54. A few easy examples from physics in a general Riemannian *N*-space with general coordinates

1. From section 39: The *distance* between two points $x^r(t_1)$ and $x^r(t_2)$ on a curve $x^r = x^r(t)$, where t is a parameter, is

$$s = \int_{t_1}^{t_2} \sqrt{g_{pq} \frac{dx^p}{dt} \frac{dx^q}{dt}} dt$$

2. The velocity of a moving point is $\frac{dx^k}{dt}$ in the coordinate system x^k. In the coordinate system $\overline{x}^j$ the velocity is $\frac{d\overline{x}^j}{dt}$. By the chain rule

$$\frac{d\overline{x}^j}{dt} = \frac{d\overline{x}^j}{dx^k} \frac{dx^k}{dt}$$

It follows that the contravariant velocity vector for a moving point is

$$v^k = \frac{dx^k}{dt}$$

The covariant velocity vector is given by $v_r = g_{rs} v^s$.

3. In a general Riemannian 3-space with general coordinates x^r, the kinetic energy T of a particle of constant mass M moving with velocity having magnitude v is given by the invariant

$$T = \frac{1}{2}Mv^2 = \frac{1}{2}M\left(\frac{ds}{dt}\right)^2$$

$$= \frac{1}{2}M\left(\sqrt{g_{pq}\frac{dx^p}{dt}\frac{dx^q}{dt}}\right)^2$$

$$= \frac{1}{2}Mg_{pq}\frac{dx^p}{dt}\frac{dx^q}{dt}$$

4. The contravariant acceleration vector of a moving point is

$$a^k = \frac{\delta v^k}{\delta t} = v^k;q\frac{dx^q}{dt}$$

$$= \left(\frac{\partial v^k}{\partial x^q} + \begin{Bmatrix} k \\ qs \end{Bmatrix} v^s\right)\frac{dx^q}{dt}$$

$$= \frac{\partial v^k}{\partial x^q}\frac{dx^q}{dt} + \begin{Bmatrix} k \\ qs \end{Bmatrix} v^s \frac{dx^q}{dt}$$

$$= \frac{dv^k}{dt} + \begin{Bmatrix} k \\ qs \end{Bmatrix} v^s \frac{dx^q}{dt}$$

$$= \frac{d^2x^k}{dt^2} + \begin{Bmatrix} k \\ qp \end{Bmatrix} v^p \frac{dx^q}{dt}$$

$$= \frac{d^2x^k}{dt^2} + \begin{Bmatrix} k \\ pq \end{Bmatrix} \frac{dx^p}{dt}\frac{dx^q}{dt}$$

The covariant acceleration vector is given by $a_r = g_{rs}a^s$.

5. Assume the mass M of the particle to be an invariant. Then $Ma^k = F^k$. F^k is the contravariant force vector, and Newton's law can be written

$$F^k = Ma^k = M\frac{\delta v^k}{\delta t}$$

The covariant force vector is defined by the equation

$$F_r dx^r = dW,$$

where dW is the work done by the forces in an arbitrary infinitesimal displacement dx^r. Remember that $F^r = g^{rs}F_s$ and $F_s = g_{rs}F^r$.

6. Let ϕ denote a scalar or invariant (tensor of rank zero). Under the transformation $x^k = x^k\left(\bar{x}^1, \bar{x}^2, \ldots, \bar{x}^N\right)$, ϕ is a function of x^k and hence $\bar{x}^j$ such that $\phi\left(x^1, x^2, \ldots, x^N\right) = \bar{\phi}\left(\bar{x}^1, \bar{x}^2, \ldots, \bar{x}^N\right)$. From calculus (by the chain rule for partial differentiation)

$$\frac{\partial\bar{\phi}}{\partial\bar{x}^j} = \frac{\partial\phi}{\partial\bar{x}^j} = \frac{\partial\phi}{\partial x^k}\frac{\partial x^k}{\partial\bar{x}^j} = \frac{\partial x^k}{\partial\bar{x}^j}\frac{\partial\phi}{\partial x^k}$$

So $\dfrac{\partial\phi}{\partial x^k}$ is a covariant vector (a covariant tensor of rank one). The covariant gradient tensor is $\mathrm{grad}\,\phi = \dfrac{\partial\phi}{\partial x^k}$.

7. If $V\left(x^1, x^2, \ldots, x^N\right)$ is the potential energy of a particle, then $-\dfrac{\partial V}{\partial x^k} = F_k$, where F_k is the covariant force vector acting on the particle. Here V is an invariant or scalar. Remember that a tensor when multiplied by an invariant is again a tensor of the same rank and type.

8. In the Riemannian 4-space of Natural Space-Time, the motion of a particle, under the action only of inertia and gravitation, is described by the equations

$$\frac{d^2x^r}{ds^2} + \begin{Bmatrix} r \\ pq \end{Bmatrix}\frac{dx^p}{ds}\frac{dx^q}{ds} = 0$$

Note that this particle moves in a straight line (a geodesic line) in this particular Riemannian 4-space. The matter and energy that are present in this 4-space determine the metric tensor g_{pq} at each point of the space, and therefore the geometry at each such point.

9. Also in the 4-dimensional Riemannian space of physics the geodesic equations reduce to that of straight lines if $\begin{Bmatrix} r \\ pq \end{Bmatrix} = 0$ throughout the space:

$$\frac{d^2x^r}{ds^2} = 0 \quad \Rightarrow \quad \frac{dx^r}{ds} = a \quad \Rightarrow \quad x^r = as + b$$

$\begin{Bmatrix} r \\ pq \end{Bmatrix} = 0$ throughout the space is a sufficient condition that the space is Euclidean. The $\begin{Bmatrix} r \\ pq \end{Bmatrix}$ are the "components of the gravitational field."

The g_{pq} field is related to the distribution of matter. Geometry becomes an expression of the gravitational field. The behavior of clocks and light is determined by the g_{pq}. Remember that the g_{pq} are functions of the coordinates and that the g_{pq} transform in such a way that ds^2 remains invariant. The light ray is not a geodesic line in three-dimensional space. Only in the 4-dimensional world is the world line of the light ray geodesic.

"Why can't you travel faster than light? The reason you can't go faster than the speed of light is that you can't go slower. Everything, including you, is always moving at the speed of light. How can you be moving if you are at rest in a chair? You are moving through time." L. P. Epstein in "Relativity Visualized," the gold nugget of relativity books.

"To deny the ether is ultimately to assume that empty space has no physical quantities whatever. The fundamental facts of mechanics do not harmonize with this view. [...] The ether of the general theory of relativity is a medium which is itself devoid of *all* mechanical and kinematical qualities, but helps to determine mechanical (and electromagnetic) events." Albert Einstein in "Sidelights on Relativity."

55. A few easy exercises

1. Write the terms in each of the following indicated sums.

$$ds^2 = g_{pq} dx^p dx^q, \quad N = 4.$$

$$x^p x^p, \quad N = 3.$$

$$a_{pq} x^q, \quad N = 5.$$

2. Write the law of transformation for the tensor A^p_{qr}.

3. Show that $\dfrac{\partial A_p}{\partial x^q}$ is not a tensor even though A_p is a tensor.

4. Show that $\dfrac{\partial x^p}{\partial x^q} = \delta^p_q$.

5. Show that $\dfrac{\partial x^p}{\partial \overline{x}^q}\dfrac{\partial \overline{x}^q}{\partial x^r} = \delta^p_r$.

6. If $\overline{A}^p = \dfrac{\partial \overline{x}^p}{\partial x^q} A^q$ prove that $A^q = \dfrac{\partial x^q}{\partial \overline{x}^p} \overline{A}^p$.

7. If A^p and B^p are tensors, show that their sum and difference are tensors.

8. Show that the contraction of the tensor A^p_q is a scalar or invariant.

9. Show that ANY inner product of the tensors A^p_r and B^{qs}_t is a tensor of rank three.

10. If $X(p,q,r)$ is a quantity such that $X(p,q,r) B^{qn}_r = 0$ for an arbitrary tensor B^{qn}_r, show that $X(p,q,r) \equiv 0$.

11. Show that EVERY tensor can be expressed as the sum of two tensors, one of which is symmetric and the other skew-symmetric in a pair of covariant or contravariant indices.

12. If $\Phi = a_{jk} A^j A^k$ show that we can always write $\Phi = b_{jk} A^j A^k$ where b_{jk} is symmetric.

13. If $ds^2 = g_{jk} dx^j dx^k$ is an invariant, show that g_{jk} is a symmetric covariant tensor of rank two.

14. Determine the metric tensor in spherical coordinates.

15. Prove that g^{jk} is a symmetric contravariant tensor of rank two.

16. If $A_j = g_{jk} A^k$, show that $A^k = g^{jk} A_j$.

17. Show that $L^2 = g_{jk} A^j A^k$ is an invariant, and that $L^2 = g^{jk} A_j A_k$.

18. If A^j and B^k are vectors, show that $g_{jk} A^j B^k$ is an invariant.

19. Determine the Christoffel symbols of the second kind in cylindrical coordinates.

20. Prove that the covariant derivatives of g_{jk}, g^{jk}, and δ_k^j are zero.

21. Show that $\left(g_{jk} A_n^{km}\right)_{;q} = g_{jk} A_{n\;;q}^{km}$.

22. Show that the intrinsic derivatives of g_{jk}, g^{jk}, and δ_k^j are zero.

23. Show that the geodesics on a plane are straight lines.

24. Show that the geodesics on a sphere are arcs of great circles.

56. Index (The numbers refer to sections)

57. Selected Bibliography

Bishop, R. L. and Goldberg, S. I. (1968). **Tensor analysis on manifolds.** Macmillan, London.

Eddington, A (1923). **The mathematical theory of relativity.** Cambridge University Press.

Eisenhart, L. P. (1926). **Riemannian geometry.** Princeton University Press.
Byron, F. W. and Fuller, R.W. (1969, 1992). **Mathematics of classical and quantum physics.** Dover, New York.

Kay, D. C. (1988). **Tensor Calculus.** McGraw-Hill, New York.

Klingenberg, W. (1982). **Riemannian geometry.** Walter de Gruyter, New York.

Lawden, D. F. (1982). **An introduction to tensor calculus, relativity and cosmology.** Wiley, New York.

Lebedev, L. P. and Cloud, M. J. (2003). **Tensor Analysis.** World Scientific Pub Co Inc.

Levi-Civita, T. (1927, 1977). **The absolute differential calculus.** Dover. New York.

Misner, C. W., Thorne, K. S., and Wheeler, J A. (1973). **Gravitation.** Freeman, San Francisco.

Nelson, E. (1974). **Tensor Analysis.** Princeton University Press, Princeton.

O'Neal, B. (1983). **Semi-Riemannian geometry: with application to relativity.** Academic Press, New York.

Parker, L. and Christensen, S. M. (1994). **MathTensor: A System for Doing Tensor Analysis by Computer.** Addison-Wesley, New York.

Rainich, G. Y. (1950). **Mathematics of relativity.** Wiley, New York.

Sachs, R. K. (1977). **General relativity for mathematicians.** Springer, Berlin.

Schouten, J. A. (1954). **Ricci-calculus.** Springer, Berlin.

Schutz, B. F. (1985). **Geometrical methods in mathematical physics.** Cambridge University Presss.

Sokolnikoff, I. S. (1964). **Tensor Analysis.** Wiley, New York.

Simmonds, J. G. (1997). **A Brief on Tensor Analysis.** Springer, Berlin.

Spain, B. (2003). **Tensor Calculus: A Concise Course.** Dover, New York

Spiegel, M. R. (1959, 1974). **Vector analysis and an introduction to tensor analysis.** McGraw-Hill, New York

Spivak, M. (1965). **Calculus on manifolds.** Benjamin, New York.

Synge, J. L. and Schild, A. (1949). **Tensor Calculus.** University of Toronto Press.

Wills, A. P. (1958). **Vector Analysis With an Introduction to Tensor Analysis.** Dover, New York.

Wrede, R. C. (1972). **Introduction to Vector and Tensor Analysis.** Dover, New York.

978-0-595-35694-2
0-595-35694-X

Made in the USA
Lexington, KY
26 August 2012